Postenvironmentalism

Chiara Certomà

Postenvironmentalism

A Material Semiotic Perspective on Living Spaces

Chiara Certomà
Scuola Superiore Sant'Anna
Pisa, Italy

ISBN 978-1-137-50789-1   ISBN 978-1-137-50790-7 (eBook)
DOI 10.1057/978-1-137-50790-7

Library of Congress Control Number: 2016948582

Cover illustration: Cover pattern © Melisa Hasan

Printed on acid-free paper

This Palgrave Pivot imprint is published by Springer Nature
The registered company is Nature America Inc.
The registered company address is: 1 New York Plaza, New York, NY 10004, U.S.A.

To Allegra and Marco

CREDITS AND ACKNOWLEDGMENTS

This book is inspired by the work of environmentalist collectives that make their best, every day, to turn ordinary places into "living spaces."

I am greatly indebted to Juliet Jane Fall for the friendly encouragement and the insightful suggestions. For sharing with me the joy of thinking through environmental politics, and the difficult task of making things public, I want to thank Eloisa Cristiani and Marco Frey, together with Barbara Henry, who supported my very first steps into the post-environmentalist theory. I had the pleasure of working with them at the Scuola Superiore Sant'Anna, which offered me the opportunity to conduct my research in an intellectually vibrant community.

In developing my work, I have been very much inspired by conversations with Christof Mauch and Helmut Thrischler at the Rachel Carson Center for Environment and Society, where I experienced the pleasure of enjoying high-level research in environmental humanities. Steve Hinchliffe, Doreen Massey, Nick Bingham, and the colleagues at The Open University introduced me to the world of material semiotics, by challenging all my previous beliefs about environmental politics.

I explored most of the topics addressed in this book with enthusiastic collaboration with Luca Colombo, Pascal Acot, Francesco Rizzi, Massimo Battaglia, Silvia Cioli, and Luca D'Eusebio; they provided me with fresh perspectives and cases to work upon.

The series editors Sara Doskow and Chris Robinson at Palgrave Macmillan have been extremely supportive in the entire managing process, together with the editors Brenda Black and Kirsten Saliste. Thanks to all. Pictures 1.2 and 5.3 have been kindly offered by my friend Alessandro Pozzi and by Zappata

Romana, while picture 6.1 is an art piece by my daughter Allegra Guerrazzi. Special thanks to all of them.

Moreover, I'm lucky to share part of my life with special people who have taught me what it practically means "to go beyond the end." Particularly, I owe a lot to my friends at the environmental associations ASUD. *Ecologia e Cooperazione,* and *Legambiente* and to the communard companions from *La Comune di Bagnaia* (and its friends). Last but not least, I thank my family for its unconditional support.

CONTENTS

LIST OF ABBREVIATIONS

ANT	Actor-Network Theory
BES	Biodiversity and Ecosystem Services
CERs	Certified Emission Reductions
CGY	Certificado de Garantía Yasuní (Yasuní Guarantee Certificates)
CSR	Corporate Social Responsibility
EMS	Environmental Management System
ICLEI	International Council for Local Environmental Initiatives
IPCC	International Panel on Climate Change
ISO	International Organization for Standardization
ITT	Ishpingo-Tambococha-Tiputini
IUCN	International Union for Conservation of Nature
TEEB	The Economics of Ecosystems and Biodiversity
UNDP	United Nations Development Programme
UNEP	United Nations Environmental Programme
UNESCO	United Nations Educational, Scientific and Cultural Organization
UNFCCC	United Nations Framework Convention on Climate Change
UN-REDD	United Nations Reducing Emission from Deforestation and Degradation Programme

LIST OF FIGURES

LIST OF TABLES

LIST OF BOXES

CHAPTER 1

Introduction

Abstract Since its very beginning in the nineteenth century, environmental thinking was characterized by the presence of different interpretations of the relationship between society and nature. The emergence of scientific environmentalism in the 1970s was welcomed as a synthesis bringing together scientific and social perspectives. As a consequence, in a few years environmental thinking reached the peak of public interest giving rise to the green diplomacy of UN summits, grassroots commitment, green political theory and green-business managerialism. However, while the progressive mainstreaming of environmental thinking attracted the most disparate supporters, it also slipped out of environmentalists' hands, and determined its progressive de-politicization—up to a seemingly death. Is this the very fate of environmentalism?

Keywords Scientific environmentalism · green diplomacy · green managerialism · environmental thinking

This book begins from an unusual point, as it does not start from the beginning but rather from the end of its object of investigation, notably from the (seeming) death of environmentalism. Despite odds, the choice is not unjustified. In fact, most of existing books on the topic, exploring the origin and development of environmental thinking, leave the reader with the uncomfortable feeling of its progressive disappearance from the

© The Author(s) 2016
C. Certomà, *Postenvironmentalism*,
DOI 10.1057/978-1-137-50790-7_1

public arena, or claim that it now firmly sits at the high table of twentieth-century political ideologies.

None of the two options, however, justify people's engagement in the continuous and hard battles environmentalisms are still fighting. Even today, we can see many people who embark on endless struggles for transforming ecological and social practices they repute to be harmful for the life on the Earth. What is inspiring them?

Through a critical analysis of the theoretical path that led environmental thinking and environmentalists close to extinction (gripped in the gearwheels of progressive normalization and the prophecy of its ineffectiveness), this book comes to disclose the emerging worldviews, ideals, and means which are now rescuing environmentalism from its own end.

Chapters 2 and 3 describe competing approaches to environmentalism and post-environmentalist theory, while Chapters 4 and 5 advance an innovative understanding on the future of environmental thinking by moving from the analysis of path-breaking interpretations of the power of networking and mobilizing beings and things in the production of ordinary (and extraordinary) environmental struggles. These chapters include, for instance, everyday initiatives of resilient planning or urban gardening, innovative transport behaviors, and also major programs supporting smart energy grids or large ecosystem restoration projects. While not (necessary) entailing brand-new practices, they all adopt a new perspective on the formation and functioning of sociopolitical collectives involved in a disputed state of affair. From such a perspective, environmental issues are reconsidered under an alternative light as they show unexpected links and connections, alliances between humans and nonhumans, overlapping of technologies and procedures, integration of cultural facts, and matter of things in multilayered public arenas, where a controversial topic is debated through practical engagement further than discursive practices.

Building upon current transformation of the relationship between science, technology, society, and the environment, this book suggests adopting the postmodern material-semiotic approach (whose most popular form is the actor-network theory (ANT); Latour, 2005) in order to appreciate the main trends in the evolution of environmental thinking. The following pages thus combine a theory-informed presentation of worldwide cases and crucial events in the history of environmentalism with a journey into scholarly explorations in order to answer the crucial question: where is environmental thinking heading toward?

1.1   A World in Commotion

Environmental thinking emerged in the Western world in the 1960s as a structured form of collective elaboration on the possibility to preserve or restore natural equilibria, ecosystem functioning, and the relationship between humans and the environment. Some have described its rise as an extension of the Western Romantic tradition, combining in a complex and somewhat contradictory form, a faith in the possibilities offered by scientific knowledge and technical innovations alongside a belief in the intrinsic value of nature. The internal conflict between the more positivist and rationalist approaches, which produced both science-based blueprints for survival and anti-scientific or even spiritual approaches, based on eco-centric and biocentric ethical perspectives (Bartollomei 1995), characterizes the panorama of environmental thinking since its inception. This dual character of environmental thinking dates back to the diffusion of colonial sciences, which promoted the rapid diffusion of new scientific ideas and, at the same time, incorporated the pantheistic thought of the colonized population in a romantic interpretation of nature so that "[t]he ideological and scientific content of early colonial conservationism . . . amounted by the 1850s to a highly heterogeneous mixture of indigenous, Romantic, Orientalist and other elements" (Grove 1995, 2). The desire to know natural laws by intimately participating in nature's spiritual richness expressed by transcendentalist (e.g., John Muir, Henry Toureau, and Adolf Just) and romantic thinkers (most notably Jean-Jacques Rousseau, Johann Gottfried Herder, Johann Wolfgang von Goethe, and Friedrich Schiller) and the desire to exercise power over the environment are both constitutive parts of early environmental thinking.

Modern environmental thinking (also referred to as "scientific environmentalism") emerged in the Western world during the 1970s as a structured form of commitment, aimed at safeguarding the fragile global ecological equilibrium by halting or limiting the impact of human production and consumption processes. Its worldwide diffusion was due to the exponential increase in the range, scale, and seriousness of environmental problems that gave rise to a massive international mobilization pointing out the environmental side effects of the long global economic boom following World War II (including rising population, increasing energy and resource consumption, new sources and level of pollution, waste production, biodiversity erosion, etc.). The tragedy of the nuclear bomb explosions; the discovery of the damaging effects of dioxins, pesticides, fertilizers, and detergents; the chemical war in Vietnam; and similar events alerted global populations. From the end of the 1970s in Western Europe and North America, a corpus of heterogeneous scientific and social narratives helped to forge the core of our belief that the Earth's ecological balance is seriously endangered (and that we, as human beings, are mainly responsible for this), and this, in turn, gave rise to the environmentalist movement. Although both environmental thinking and the environmentalist movement included a plethora of distinct understandings of the consequences of environmental problems and the best ways for dealing with them, it can be agreed that together they all contributed to the popularization of a common wisdom. Modern environmentalism can be thus regarded as "a box that contains anarchist and protofascist, Marxist and liberal, natural scientist and visionary alike; not because their world-views were identical, but because all shared an idea which by them was perceived as primary, although its secondary manifestations may have differed" (Bramwell 1989, 237–238).

The most prominent aspect of modern environmentalism—which deeply differentiates it from other sociopolitical movements—resides in the close interrelatedness of its sociopolitical claims with scientific findings and its dependence on scientists' advice, so it has been often referred to as "scientific environmentalism." Particularly the ecological paradigm of Ecosystem Ecology outlined by Eugene Odum, George Hutkinson, and Ramon Margalef (1977), together with the Global Ecology theory (the so-called Gaia hypothesis) elaborated by Jim Lovelock (1979) decisively contributed to the establishment of ecology as a unified and authoritative scientific discipline from the late 1970s onward, backing up on the widespread diffusion of the Ludwig von Bertalanffy's General Systems Theory (Certomà 2006). Moreover, ecosystem and global ecologists greatly contributed to the diffusion of basic concepts of ecology (e.g., ecosystem, natural balance, carrying capacity, etc.) as well as raising public awareness on the urgency of global action for ecological protection.

A wide range of best-selling books made the general public familiar with some of the most pressing dangers the global population was going to face because of the heedless alteration of ecological cycles and the erasure of nature's capability to react and counterbalance anthropic stresses. Among them, some need to be mentioned for the large impact they had on both public and governments; these include the *The Limits to Growth* report (Meadows et al. 1972), which predicted an impending ecological catastrophe unless exponential economic growth were to be replaced with "steady-state" economic development strategy; *Silent Spring*

(Rachel Carson 1962), which described the toxic effects that pesticides have on humans, environment, and animals and which led to the global ban of the insecticide DDT; and *The Population Bomb* (Ehrlich 1968), which warned about the danger of human mass starvation due to over-population. Together with publications and campaigns describing the ecological and social consequences of Western modern lifestyle, a number of blueprints for survival that suggested new paths for achieving sustain-ability and restoring altered equilibrium also became largely popular (e.g., Barry Commoner's *The Closing Cycle* (1971), Ernst Schumacher's *Small Is Beautiful* (1973), and Arne Naess' *Ecology, Community and Lifestyle* (1989)).

An important role was played by the emerging environmental associa-tions, NGOs, and eco-communities that gathered ordinary citizens, practi-tioners, and scientists together around the cause of saving the planet from ecological disaster. A number of largely diversified associations were estab-lished, ranging from the radical environmentalism of *Green Pirates* (Taylor, 2013) and the so-called eco-terrorists of the *Earth Liberation Front*, to the hippie approach of the *Global Ecovillages Network*, the scientific environ-mentalism of large NGOs such as *Greenpeace* or *Friends of the Earth*, to the more conservative struggle for nature preservation advanced by the *WWF*, and a plethora of initiatives aimed at promoting collective environmental friendly behaviors. Most of them became relevant actors in the interna-tional sociopolitical arena, entering the debate on a number of environ-mental issues and related matters, such as the unsustainable consequences of global free-trade economy, the bad working conditions in the poorest countries of the world, and the progressive loss of natural habitats, biodi-versity, and cultural traditions worldwide.

The broad mobilization of energies and ideas found its momentum in the first *UN Conference on the Human Environment* held in Stockholm in 1972. It was during the Stockholm Conference that international dignitaries first officially recognized that the ecological stability of the Earth is threatened by anthropic activity and worked collectively toward defining universal values and developing globally accepted rules and blueprints for global ecosystem protec-tion. Environmental concerns, defined by the Stockholm conveners as "dan-gerous levels of pollution in water, air, earth and living beings; major and undesirable disturbances to the ecological balance of the biosphere; destruc-tion and depletion of irreplaceable resources; and gross deficiencies, harmful to the physical, mental and social health of man [*sic*]," (UNEP 1972, 3) entered the international debate and rapidly climbed up the ladder of shared priorities to gain a prominent role in the global political agenda. Some years later the conference produced the seminal report *Our Common Future* (also called *Brundtland Report*, UN World Commission on Environment and Development 1987) elaborated by the newly established World Commission on Environment and Development (1987) and lead to creation of the United Nations Environmental Programme (UNEP). This introduced the groundbreaking and heavily debated concept of "sustainable development" and started a process of environmental thinking mainstreaming in the context of global culture. At the same time, however, participants of the Stockholm Conference, in the attempt to find an agreement between Northern and Southern countries' understandings of what a blueprint for survival should be, sowed the seeds of future disagreements by claiming that it is possible to decouple economic growth and pollution and, thus, to simultaneously achieve economic well-being and environmental sustainability. In so doing, the official documents issued from that moment onward by international organizations gave cause for reflection to activists, policy planners, and scholars worldwide, who wondered how the apparent oxymoron of sustainable development can be justified by alternative interpretations of development, environment, and growth. Despite the broad consensus the concept has been able to attract, in fact, its generality and ambiguity have made it very difficult to be operationalized (Berke and Conroy 2000); thus, sustainable development has become a less fashionable expression today: "The vacuity of the way it is so often used as a euphemism for growth for its own sake has become widely known. Environmentalists never really liked the phrase, but they took advantage of its endorsement by the establishment to start talking more and more about 'sustainability'" (Dresner 2002, 80). As a consequence, a myriad of alternative, more nuanced interpretations of the *Brundtland Report's* words emerged in time (Mebratu 1998) and gave rise to diverse courses of action, from weak to strong sustainability models (Neumayer 2010) in both social economic and political domains (Rees 1995).

1.2   GREEN, GREEN, MY WORLD IS GREEN...

In the 1980s, environmental thinking reached the peak of public interest and attention as it entered the international political arena and significantly influenced the regional production of innovative policy measures for global environment protection and sustainable development under the auspices of a large number of dedicated UN summits (most notably the Rio de Janeiro 1992 "Earth Summit," the

Johannesburg 2002 summit, and the 2012 Rio+20 summit) reaffirming the primacy of the sustainable development ideal. Thousands of declarations and protocols encompassing virtually all aspects of the relationship between society and environment (climate change, biodiversity loss, desertification, and much more) expanded the field of competence of the nascent fields of international environmental diplomacy and environmental politics. Green diplomacy entered its golden age with the proliferation of international meetings and subsequent documents dealing with a wide range of problems affecting the Earth's ecological balance (e.g., *The World Conservation Strategy* by IUCN, UNEP, and WWF in 1980; The European Conference on Sustainable Cities and Towns that resulted in the Charter of European Cities and Towns Towards Sustainability, also known as the Aalborg Charter in 1994; the Kyoto Protocol issued by the UN Framework Convention on Climate Change in 1997; and the Bali Climate Change Conference in 2007). At the same time, the work of newly established research institutes for environment and society (e.g., the Wuppertal Institut für Klima, Umwelt, Energie founded in 1991) produced evidences that confirmed the importance of these endeavors.

Scientific research showed that the Earth's biodiversity and natural resources were progressively decreasing, and these problems became common knowledge; international and national policy measures for avoiding global environmental disaster greatly expanded, and even the tiniest gestures of everyday life—such as choosing tomatoes in the supermarket, changing a light bulb, or turning off the water tap while brushing teeth—acquired a new sociopolitical significance. New landmark principles for international action toward environmental protection have been introduced in both official declarations and ordinary communication, including the concept of carrying capacity and the ecological footprint (Rees and Wackernagel 1994), the principles of precaution (Gollier et al. 2000), the common but differentiated responsibility (UN Conference on Environment and Development, 1992), and the environmental right framework (Boyle et al. 2009). Those were largely popularized by new dedicated media outlets such as the leading British journal *The Ecologists* (founded in 1970 and ended in 2009), or the Italian journal of scientific environmentalism called *La nuova Ecologia* (The new ecology, founded in 1980), closely followed by dedicated scientific reviews spanning nearly the entire widths of the disciplinary spectrum (e.g., *Environmental Politics* or *Capitalism, Nature and Socialism*) (see Fig. 1.1).

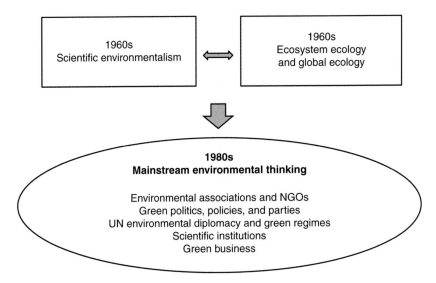

Fig. 1.1   The transformation of environmental thinking

Source: The author

The pervasiveness of environmental issues in all the spheres of social and political intervention determined to the emergence of specific environmental regimes, i.e., distinct institutions (including laws, apparatuses, framework treaties, and regulatory procedures) dealing with specific environmental issues (de Vos et al. 2013) by adopting systems of rights and obligations and related decision-making procedures in international environmental policy (Biermann 2006; Kates et al. 2005). Most of the policy measures that have been undertaken reached across the boundaries of different political theories and alternatively inspired liberal (Wissemburg 1997) or socialist approaches (O'Connor 1998), depending on how far national governments or regional institutions were prepared to restrict individual freedom for the sake of pursuing sustainability. At the same time, the emerging green political theory advanced a critique of both Western capitalism and Soviet-style communism as essentially two different versions of the same overarching ideology of industrialism—despite their differences concerning the respective roles of the market and the state.

The emergence in the late 1970s of the Coordination of European Green and Radical Parties[1] all over the world saw environmentalists entering many national parliaments and negotiating their values with the representatives of supposedly antagonistic interests. Political philosopher Andrew Dobson in fact asserts that "understanding the political and intellectual nature of Green politics means seeing that its political prescriptions are fundamentally left-liberal, and if a text, a speech or an interview on the politics of the environment sounds different from that then it is not green but something else" (Dobson 1990, 83–85). In practice, Western democracies tended to combine environmental values with liberalism. This can be understood as a result of the attempt of different political parties to endorse *en vogue* green principles and make them fit into their political program, as they were able to attract broad consensus. For instance, in addressing the Scottish Conservative Party Conference in February 1982 about the Falkland War, UK Prime Minister Margaret Thatcher listed environmental issues as merely one of the humdrum issues she had to deal with (Thatcher 1982). However, only 4 years later, in a speech to the Royal Society, Thatcher seemed to have changed her mind and affirmed that "protecting this balance of nature is therefore one of the great challenges of the late Twentieth Century" (Thatcher 1988); she suggested that by knowing the cause of environmental problems, the government would be able to find appropriate solutions for sustainable economic development because, as Thatcher herself claimed, stable prosperity can only be achieved when the environment is safeguarded (McCormick 1991). Did Margaret Thatcher become an environmentalist? History says this was not the case; she did not change, but environmental thinking did, as it became part of Western governmental strategies.

Since then the environmental cause attracted the most disparate supporters, including Islamic fundamentalist leader Osama Bin Laden, who in his speeches is reported to have railed against industrialized nations for their responsibilities in climate change and nature destruction, and US President Barack Obama, whose ambition was to rescue the American economy by increasing green energy production (White House 2014). Yet despite the massive commitment from world leaders, extensive research conducted in determining the causes and solutions of some of the most pressing global environmental problems (e.g., the reporting activities of the Intergovernmental Panel on Climate Change (IPCC), awarded the Nobel Peace Prize in 2007), and the introduction of environmentally friendly tools and procedures across the globe (spanning from the diffusion of compostable plastic shopping bags to the UNFCCC Emission Trading Scheme

Fig. 1.2   An environmentalist protest against TEBIO congress on biotechnology, Genua, May 2000. Banner says: "Quando il mondo è in vendita, ribellarsi è naturale" (When the world is on sale, rebellion is natural)

Source: Alessandro Pozzi

negotiated in Kyoto in 1997), scientific data, and the global cultural imaginary have persisted in describing ecological decline as the inevitable consequence of people's collective environmental negligence, feeding catastrophic communal fantasies of pollution, life erosion, and degradation for the future of the planet (see Fig. 1.2).

In the late 1980s and 1990s, environmentalists believed that people's support was a positive sign of a collective green awaking, but they actually underestimated the effect of this popularization on environmental thinking itself. The emergence of the managerial attitude adopted by international organizations and large NGOs, according to which the solution of environmental issues was a matter of scientific and technical development or juridical creation of rights and enforcement of procedural norms, contributed to its transformation. Having being once so innovative and influential, environmental organizations and green political parties soon faced continuous failure and lost much of their appeal to the general public. Today, the attention given to environmental issues is no longer accompanied by a strengthening of environmental culture or any related political movements. While a number of challenging issues climbed to the top of public interest (and there remained for decades)—including air and water quality and chemical pollution, the growing amount of waste, and the depletion of natural resources (as the Eurobarometer data by European Commission clearly show)—this period nonetheless also signaled the beginning of the decline of environmentalism. Many social scientists noticed, than, that when this happened, environmental thinking changed its face by limiting itself to survival strategies (Castree 2006) and progressively lost its appeal for the wide public.

Box 1.1 Communication and de-politicization: Al Gore's "convenient" strategy about global warming
Political scientist Timothy Luke suggests that the green geopolitics of US President Bill Clinton administration is an excellent example of the attempt to integrate environmental protection into national economic plans and international diplomacy:

> The environment, particularly the goals of its protection in terms of 'safety' or 'security', has become a key theme of many political operations, economic interventions and ideological campaigns to raise public standards of collective morality, personal responsibility and collective vigour. [E]nvironmental protection issues, ranging from resource conservation to sustainable development to ecosystem restoration, are getting greater consideration in the name of creating jobs, maintaining growth, or advancing technological development (Luke 1998, 122–123).

Former vice president Al Gore's commitment toward climate change issue is a particularly clear example of how national governments can appropriate the language of environmentalism "while advancing a specifically liberal kind of environmentalism, ... with its emphasis on markets, eco-taxes, the exercise of consumer conscience, and voluntary codes for firms" (Castree 2006, 14).
While advocating for a change of behaviors in both individual and collective life, he reaffirmed the possibility to link sustainability-oriented behavior with longstanding economic and industrial interests

of the United States. In 2007, Al Gore, jointly with the International Panel on Climate Change, was awarded the Nobel Peace Prize "for their efforts to build up and disseminate greater knowledge about man-made climate change, and to lay the foundations for the measures that are needed to counteract such change" (The Nobel Peace Prize 2007). His documentary *An Inconvenient Truth* (2006) explains the causes and the foreseeable consequences of global warming and offers proposals for addressing these challenges and effectively counteracting climate change. While Al Gore's environmental commitment— as he himself presents it in the documentary—has deep roots in his personal history, after receiving the Nobel Prize he presented his message of warning and hope around the world mainly by means of keynote speeches, his movie, and best-selling books (The Assault on Reason, 2008; Our Choice, 2009). Gore's version of environmental thinking moves away from blaming of Western materialism and consumerism and turns toward a pantheistic sort of claim for a new human-nature relationship (Luke 1998). The most critical issue for individuals and society is, according to Gore, facing the inconvenient truth that the US way of energy consumption is severely endangering the planet and thus needs to be seriously reconsidered. If environmentally friendly technologies, regulatory procedures, and education programs are integrated into US politics and citizens' daily life, they can help to increase the collective environmental awareness and to fix the problems. In achieving this, communication processes are crucial—it is no coincidence that Gore's documentary devotes much attention to unveiling the inconvenient truths hidden by slanted media reports on climate change. Information based on serious scientific results and education programs, together with dedicated communication campaigns, Gore argues, should aim to attract convergent and bipartisan consensus about the urgency of environmental issues. By adopting a non-partisan approach, Gore translates the climate and energy issue into a challenge that can be faced in the context of the national economy and culture. He states, in fact, that "the market will work to solve this problem if we can accomplish this. Help with the mass persuasion campaign that will start this spring. We have to change the minds of the American people. Because presently the politicians do not have permission to do what needs to be done" (Gore 2006). Gore's speeches aimed to de-politicize environmental claims so as to make them acceptable for those actors whose green pedigree is questionable but whose economic power allows them to set the new standard of acceptability of environmental discourses.

Environmental issues progressively slipped out of environmentalists' hands as they were unable to take advantage of their suddenly procured power for pursuing their cause; at the same time, other actors with very few green credentials (e.g., multinational companies) appropriated environmental discourses, praising the virtues of green managerialism (Castree 2006) and introducing environmental values in their corporate social responsibility (CSR) principles. A business-compatible commitment to nature and ecosystem preservation become quite a popular approach for private companies to recreate their image as green and significantly change their marketing strategies (accompanied by sometimes only minor changes in their production and distribution models). This is based upon a strongly positivist attitude toward scientific progress and an appreciation of natural capital mostly as a finite source of raw material (or, in some cases, a receptor of wastes) that needs to be preserved for the invaluable services it is able to provide to humans—including those science is still unaware of (e.g., properties of underexploited botanical chemical agents).

As environmental protection turned into a profitable investment, business lobbies pushed for legal and economic support for securing control and ownership over natural resources (Katz 1998)—a support soon provided by international diplomacy as, for example, the establishment of about 280 "partnership initiatives" during the Johannesburg summit. When businesses appropriated green values, many people questioned the honesty of their environmental commitment and criticized this as being a form of "greenwashing," which creates the illusion that an otherwise dirty company is a clean one by diverting consumers' attention to minor acts meant to demonstrate environmental conscience.

Box 1.2 Dealing with environmental issues through corporate social responsibility: the UN Global Compact
In a contribute published in the review *Business Horizon* in 2012, Martin McCrory and Kyle Langvardt (2012) pointed out how a

serious and non-partisan discussion on the relationship between environment and business would have been impossible only two decades ago. Nowadays, environmental protection is a key component of corporate social responsibility (CSR) for most of the Fortune 50 firms (such as Apple, Google, Coca-Cola, and IBM), which devote prominent attention in their communication campaigns and social projects to environmental sustainability (McCrory and Langvardt 2012). Consumers are informed about corporations' commitment to mitigating the impact of production activities, and improve—or at least to preserve—environmental conditions through preventive or restorative programs. In recent years, an increasing number of companies worldwide have adopted an environmental code of conduct, ranging from Environmental Management Systems and ISO 14000 standards to sector-specific programs (e.g., the Fair Trade or Rainforest Alliance), or dedicated programs for reducing pollution along the production and consumption chain, and supporting virtuous behaviors (e.g., Starbucks rewards farmers who conserve and improve soil structure) (Fisher et al. 2005).

In the wake of companies' green awakening, the UN established the Global Compact, a global strategic policy initiative for businesses announced in 1999 aimed at aligning business strategies with a number of universally accepted principles concerning human rights, labor, environment, and anti-corruption. Adhesion to the Global Compact's non-binding principles is voluntary, and recent figures show an increase in the popularity of the initiative with a strong increase in the number of partners involved and a nearly equivalent percentage of partners distributed among the corporate, small- and medium-enterprises, and non-business sectors on almost all the continents. The Global Compact includes a number of key pillars, including both environmental and social issues (caring for the climate, biodiversity and ecosystems, sustainable energy for all, water, as well as gender equality, children, indigenous people, and social entrepreneurship).

The rationale for businesses' adoption of the Global Compact's environmental requirement resides in the recognition that economic prosperity and human well-being are strongly dependent upon natural capital, which is the world's stock of natural assets. Needless to say, all environmental issues are considered with respect to the possible impact on human environment, which is the life support system of this planet upon which human livelihood depends, rather than on environment per se, as the functional and structural variability in genes, species, and ecosystems and their functionality is assessed in regard to their ability to provide services. The OECD's Environmental Outlook to 2050 report (2012) claims that erosion, loss, and damage of natural capital provoke irreversible changes that deeply affect our lifestyle, have significant costs in terms both of restoration or substitution, and can also severely affect private sector performances. Viewed from one side, companies rely on goods and services provides by biodiversity and ecosystem services as inputs for production and processes; from the other side, ecosystems absorb and process the undesired outputs of production processes. For instance, the Economics of Ecosystems and Biodiversity report has estimated the annual cost of lost biodiversity and ecosystem degradation to be about US $2–4.5 trillion over a 50-year period (TEEB, 2012). This means that the sudden interest in environmental issues, in most cases, was motivated by the desire to provide one's own production and trade activities with the best conditions for the business to grow—even under adverse and uncertain socio-environmental conditions. In line with the neoliberal turn in global environmental thinking, the UN advanced the Global Compact strategy to contribute to the integration of sustainability concerns into business activities as part of corporate strategies, in order to contribute to achieving long-term profitability, as well as broadening sustainability goals. Businesses depend upon and have a direct or indirect impact, for instance, on biodiversity and ecosystem services (BES) (including goods provision, regulation of ecological processes, and nonmaterial benefits) through their operations, supply chains, or investment choices; it is thus important for businesses to integrate environmental considerations into their practices and to participate in the sustainable and equitable use and conservation of them (Winn and Pogutz 2013). The failure to manage impacts and dependencies on biodiversity services poses a wide range of risks (including operational, regulatory and legal, reputational, or market and financial risks), which can potentially affect a company's competitiveness and profitability and increase its liabilities—thus threatening its long-term viability (UN Global Compact and IUCN 2012). The loss or degradation of BES can affect a company's operations by reducing productivity, disrupting activities, or limiting access to resources, resulting in increased operating costs. In terms of regulatory risks, companies may find it difficult to secure a legal or social license to operate if they are not accountable for ecosystem management. As a result of poor environmental practices, they may also face legal or financial liabilities that can ultimately hurt a company's reputation, decreasing brand and shareholder value. Finally, clean-up and compensation costs resulting from environmental disasters and malpractices can severely affect a company's bottomline, as well as its reputation.

However, despite wishful interpretations of business as a valid assistant to communities and state in achieving sustainability goals (McCrory and Langvardt 2012), the search for ever-more sophisticated technical tools to assess and reduce companies' impact on the environment, frequently induced the suspicions that by proliferating data and measurements the business sector is just making accountability more difficult (EJOLT 2015). Moreover, although UN itself guarantees the sincerity of corporates' engagement in endorsing socio-environmental principles of the Global Compact, when choices between conflicting values need to be taken, there is serious doubt about whether businesses will choose the more environmental option.

With the aim of making their claims as shared as possible and looking for allies wherever they were, environmental thinking faced the unexpected consequence of diminishing people's enthusiasm precisely because sustainability turned into the new mantra of twenty-first century that everybody was praising, but very few were committed to.

Since the early 1990s, many theoretical critiques, including postenvironmentalism, have pointed out the weakness of the sustainability paradigm and the inefficacy of mainstream environmental politics. Together with cognate critical approaches, including post-structuralist critical theory (embraced for instance by sociologist Ulrich Beck or philosophers Robin Ekersley and Eric Darier), and postmodern theories (adopted by, among others, Val Plumwood and Bruno Latour), postenvironmentalism recognizes that from the 1980s onward, environmental summit diplomacy, business and corporations, and large NGOs, notwithstanding their apparent differences, determined in common a process of de-politicization of environmental issues in the pursuit of establishing a global environmental regime based on broad consensus. Unlike other critical approaches, post-environmentalism focuses on the internal reasons that challenged the reliability and effectiveness of environmental thinking and resulted in the current crisis of environmentalism. In so doing, it provides ways to dig deep into the past and present of the complex relationship between power, science, and the environment and invites us to step beyond the frontiers of the current environmental debate to find hints of how it may evolve in the future. The need for a dedicated analysis of post-environmentalism is suggested by the recognition that it is time to break the boundaries of sectorial interests (including environmentalists' interests) and to consider socio-environmental phenomena as a whole (as political ecology scholars, such as Wolfgang Sachs (1993) or Juan Martinez-Alier (2003), have already suggested).

Existing post-environmentalist approaches, however, while offering insightful analyses of the failure of current environmentalism, do not provide equally inspiring ideas about its future. In particular, they do not explore the new forms of participation and commitment with green values generated by the networking practices of heterogeneous actors (including humans, nonhumans, and more-than-human entities) in the public space.

The questions Chaps. 2 and 3 will address is: why did environmental thinking come to this dead end?

NOTE

1. The Coordination of European Green and Radical Parties was first organized for European Parliament elections in 1979.

CHAPTER 2

At the Edge of Environmental Thinking

Abstract From the 1980s onward, environmental concerns became part of the international political agenda with sustainable development turning into one of the pillars of contemporary sociocultural, political, and economic programs. This chapter analyses the two main approaches that made it possible, i.e., the realist and the constructivist one. The former (adopted by UN agencies, large NGOs, government, and business companies) prescribes the acquisition of as much as possible accurate and reliable data, which can provide tangible evidence of the pervasiveness of the problems. The latter is advanced by critical scholars to unveil the social construction of nature. Despite their differences, both of them grant the experts with the authority and legitimacy to combine nature, politics, and science in frameworks for action. This brings about, together with the search for a wise and efficient management of natural resources, also a number of normalized environmental discourses operating on people's opinions and behaviors.

Keywords Realism · constructivism · reflexive modernization · post-environmentalism · post-ecologism

In a short but compelling article published on the online journal *The Conversation* in 2012, Timothy Devinney (2012), a professor at the University of Technology in Sydney, insightfully explains the end of environmental thinking as a consequence of the clamorous lack of

© The Author(s) 2016
C. Certomà, *Postenvironmentalism*,
DOI 10.1057/978-1-137-50790-7_2

success of both the international summits' diplomacy and the global environmental movement. The UN Rio+20 Conference in 2012 dramatically ended with international delegates coming to no agreement worthy of mention. Global society is today facing massive environmental problems (which are, at the same time, social, economic, cultural, and political problems) too large and too widely distributed to be taken on by single initiatives—even if they might be effective and straightforward. After 50 years of noticeable victories and achievements, the major environmental organizations and associations are now unable to point out a single boogeyman which would make sense to fight against with any chance of success. Most environmental propaganda still portrays multinational corporations and businesses as the bad guys, but the apparent truth is that everybody is a potential destroyer of the environment because of common everyday choices and behaviors (and probably even the increasing human population on the Earth is problematic from an ecological perspective). Devinney concludes that, contrary to what environmentalists think, ordinary people do not act like activists. Instead, they get on with their lives and decision by ranking environmental protection lower when making daily decisions. This does not prevent them, however, from expressing a strong concern in general about climate change, water shortage, pollution, deforestation, and similar. However, similar concerns, although persistent, are rather pale and middling compared to other concerns exacerbated since the world economic crisis in 2008 (such as global security, safety, justice, welfare, or jobs).

The analysis contains a large amount of truth. Nowadays, the end of environmental thinking and the failure of environmentalism are common topics for debate. A significant number of environmental scholars have proclaimed with variable emphasis the death of environmental thinking. In some cases, their critical analyses served no other purpose than feeding the debate with words of complaint, blame, or atonement. They see the ashes but don't catch the phoenix that might resurge from them. In the rest of this book, we dig into the ashes of what have been stigmatized as the most serious lacks and faults of environmental thinking and look for the new environmental thinking that could emerge from the ruins and attract people's attention, inspire commitment, and offer engaging strategies for dealing with environmental challenges.

Even after the end, there may still be something that is worth of attention—despite not speaking loud proclamations of victory. There may be a way of thinking and acting that is changing the matter and the meaning of people's everyday relationships with their environment—and in so doing, it is transforming environmental thinking in general and environmentalist practices, too. I argue that this is a globally impacting "postenvironmentalist" thinking practiced in local places (and what is the global, after all, if not the multiplicatory effect of the overlapping of a manifold of locals?).

The description of the rise and subsequent decline of environmental thinking provided in Chap. 1 suggests that the crucial questions are: How have we come to this point marked by a sort of general resignation—if not cynicism—toward the real possibility for environmental thinking to appreciate and to fix the world's most pressing environmental concerns? Why has environmental thinking, which in the 1960s and 1970s was an inspiring and powerful sociocultural force challenging the global socio-economic system, lost much of its appeal and strength today?

2.1   REALISM AND CONSTRUCTIVISM

The analysis of the rationale environmental thinking can help us to unveil the reasons why the spread of environmentally friendly practices in people's everyday lives (such as recycling, energy-saving choices, or preference for public transport) was accompanied, paradoxically, with a decrease in enthusiasm and commitment toward the environmental cause and a loss of faith in the possibility for global environmental politics to succeed. Many believe that current global economic, social, and political crises (most notably the 2008 economic crisis, the humanitarian emergences in many poorest countries, the violence of the clash of cultures in the Middle East, etc.) diverted public attention and swallowed even the darkest green wisdoms. However, external causes are only partially responsible for the public forgetfulness of environmental disasters. Instead, to a large extent this phenomenon can be attributed to the incoherence and opposing positions characterizing the debates within environmental thinking. Two principal approaches can be detected in the broad spectrum of environment-related discourses and practices, namely the realist and the constructivist position. They present two different understandings of the *nature* of environmental issues based on two different representations of the world (i.e., two different epistemologies), and they advance different strategies for dealing with the issue they repute to be the most urgent to face.

The realist approach is exemplarily adopted by a quite large number of the most diverse actors, including international organizations (e.g., dedicated UN agencies, such as UNEP or UNDP), large environmental NGOs (e.g., the *WWF* or *Greenpeace*), and their supposed enemies, i.e., business companies whose activity has a direct impact on the environment (for instance, seed companies, like *Monsanto*, or oil companies, like *Shell*). Despite apparent differences, they all ground decisions, campaigns, and actions on scientific results—or claim to do so and "seek to justify their actions by reference to nature in itself" (Castree and MacMillan 2001, 219).

Box 2.1 Farmers' associations, Monsanto, and the seeds of discord: the Italian campaign "Grano o grane" against genetically modified wheat

Since the birth of modern scientific environmentalism, the relationship between environmental thinking and scientific results was a most complex one, as science played a crucial role in understanding the emergence and development of environmental issues. Scientific discourses are intended to represent things as they really are; they offer reliable evidence for elaborating strategies on the base of objective knowledge. Put this way, science is a powerful means to advance social, economic, and political claims legitimated by the authority of data. The plurality of scientific results and the plurality of their interpretations explain how, by adopting a realist approach, radically different actors can enter in antagonistic relationships on a common matter of concern by presenting different evidence in support of their claims. The convergence of both environmentalists and one of their enemies, i.e., multinational seed companies, on the same battleground of GMOs is a recurrent, still clear example of how contrasting realist positions can face each other in the public space. For instance, an Italian network of consumers, environmentalists, citizens, and producers' associations (in close relationship with cognate international networks) faced the giant Monsanto company when, at the end of 2002, it requested from the relevant US and Canada national offices the authorization for commercializing a new variety of GM wheat tolerant to glyphosate herbicide, i.e., Roundup Ready Wheat

(RRW). The magnitude of the impact of such an initiative was suddenly evident: for the first time a commodity destined to be used for human food would have been added to the available GM products (up to now only including commodities used for feeding livestock). Wheat is the principal agricultural commodity commercialized on the planet and the second most important food in the world, with an average global annual production ranging to about 600 billion tons and a total production surface of 200 billion hectares (Colombo 2004a). Because of wheat's particular role in the Mediterranean diet, the Italian food producers and consumers, although not immediately involved, expressed a strong opposition against the commercialization of GM wheat. As in other countries (particularly the United States and Canada), in Italy the broad mobilization of wheat producers, environmental and food associations, farmers' unions, and consumers' organizations (including Coldiretti, AssoCAP, CNA Alimentare, FLAI-CGIL, COOP, Grandi Molini Italiani) assembled around a campaign evocatively titled *Grano o Grane?* [Grain or problems?] led by the environmental foundation *Consiglio dei Diritti Genetici* [Genetic Rights Council].

Apart from the geopolitical reasons concerning production and market control that motivated both Monsanto and farmers' associations, the interesting point is that both of them built their claims on the basis of scientific data, and the entire dispute focused on producing documents intended to show the most authoritative and incontrovertible evidence of the validity of one or the other's claims.

When presenting its request, Monsanto produced numerous scientific studies, conducted both by their internal laboratory and by universities, that explained how biotechnologies have proved to be "highly effective ways to treat human disease, to manufacture chemical products, to eliminate waste, and to ensure abundant, healthful and affordable food for our world's growing population" (Monsanto 2003). These confirm, Montsanto argued, that "the Roundup Ready system in other crops is a proven, highly effective weed control tool that saves growers time and money. Years of field-trial data suggests the Roundup Ready system can offer North American wheat growers a compelling set of technical benefits" (Monsanto 2003). These benefits include broad-spectrum weed control; increased crop safety and yield for the possibility of early seeding; conservation-tillage enhancement resulting in loss soil erosion; improved crop quality; and reduced environmental risk thanks to the overall reduction of herbicide use (Monsanto 2003). Dedicated researchers showed the Roundup herbicide to be more effective in terms of ecological footprint reduction compared to other herbicides (Grenier 2002).

In order to answer the evidence offered by Monsanto, the Italian network, together with the Canadian Wheat Board, US farmers' and producers' associations and Japanese farmers (Schubert 2001), gathered anti-GM scientists from the Italian National Research Institute for Food and Nutrition, the National Institute of Agrarian Economics, and the Universities of Florence and Bari to support the *Grano o grane?* campaign with evidence and fresh data contesting those offered by Monsanto. The resulting Italian report analyzes, on the basis of both international and national research, the possible health effects of GM introduction into the human diet (e.g., allergies, intolerances, and resistance to antibiotics) (Colombo 2004b); the ecological implications (e.g., herbicide tolerance, consequences of individual management practices and cropping systems, gene spread contamination, risk to increase selection pressure for developing wheat resistant to glyphosate, and decrease in biodiversity) (Van Acker et al. 2003). Economic and social consequences were also considered, including the consumers' acceptance of the product and the financial and marketing effects of the competition between national wheat producers and the already established monopoly on GM crops by Monsanto (Gillis J. 2003).

The Italian associations of wheat producers and processors not only threatened that they would refuse to buy GM seeds but also to completely avoid importing from those countries where it is known that GM wheat was grown; as Italy is a modest producer of wheat but is one of the principal processors and consumers, the threat was taken seriously. In 2004, the executive vice president of Monsanto, Carl Casale, yielded: "As a result of our portfolio review and dialogue with wheat industry leaders, we recognize the business opportunities with Roundup Ready spring wheat are less attractive relative to Monsanto's other commercial priorities" (Monsanto, 2004). The international mobilization was thus powerful enough to influence public opinion and economic operators, and to force Monsanto to freeze its marketing plan for RRW.

Realists believe that scientific evidence not only portrays reality but also provide implicit suggestions for deducing how things *have to be* from the representation of how things *are* in their "natural state." This means that realist discourses tend to retrieve moral imperatives from scientific evidence—or more subtly, to use scientific results as a basis for legitimating political advice on socioeconomic strategies (Luke 1999).

For instance, the concept of the balance of nature that, from Aldo Leopold's *A Sand Country Almanac* (1968) onward strongly influenced the environmental conservationist campaigns (claiming that "a thing is right if it tends to preserve the integrity, stability and beauty of a biotic community. Wrong when it tends otherwise" (Leopold 1968, 224)) is based on the (supposed) scientific evidence that ecosystems tend to stability—and we should adopt this as a regulatory goal for sociopolitical interventions. Clearly, the balance of nature discourse linked together ecological paradigms and sociopolitical considerations (Trudgill 2001), and a long-lasting debate has involved those who considered stability as the main forces in the regulation of ecosystems dynamics,[1] and those who believed that chaos is the rule of ecosystems dynamic (Gleason 1939; Mayr 1997). Contrary to the common wisdom that considers disequilibrium as the effect of anthropic interferences with ecosystems,[2] the latter group claims that ecosystems do not tend to equilibrium and their chaotic functioning "can provide no absolute legitimation for particular scientific positions, and science in that sense can provide no legitimation to which a particular politics can appeal" (Massey 2006, 43–44). It is, thus, obvious that data are not self-evident; they are selected and interpreted, then used to legitimate one or another course of action. Some apparent contradictory outcomes may arise from this. For instance, the same scientific results lead UNEP to recognize the alarming truth that the Earth has finite carrying capacity, and also to advance the more optimistic suggestion that its upper limits can, however, be expanded by adopting sophisticated technological, social, and economic means of intervention (Pengra 2012).

From the 1990s onward, in opposing what they reputed to be the *naïve* realist view, many social scientists and most notably Critical Thinkers (sociologists, anthropologists, and philosophers building upon the legacy of Frankfurt School) advanced an alternative understanding of environmental issues, which is intended to reveal the social construction of nature and is generally referred to as constructivism. They claim that technical hieroglyphs have been superimposed on people's perceptions, and both environmentalists and their industrial opponents are prone to realistic fallacy:

> The observable consequence is that critics frequently argue more scientifically than the natural scientist they dispute against... [but] fall prey to a naïve realism about definition of the danger.... On the one hand, this naïve realism of the dangers is apparently necessary as an expression of outrage and motor of protest; on the others is its Achilles' heel. (Beck 1995, 60)

Constructivism relies on the belief that reality is actually the product of our categories of thought, and this means that the basis for environmental political decisions is not provided by scientific evidence but by a social imaginary that elaborates in cultural terms the available scientific results. As a consequence, environmental issues cannot be effectively addressed except by changing our understanding of nature and our sociocultural behaviors, because constructivists read "all instances of human/non-human relations as somehow culturally determined" (Hinchliffe 2003, 207). The constructivist epistemological process originates from a semiotic exploration of the relations between society and nature: the observer literally discovers separate, three-dimensional objects and creates a specific narrative to interpret them.

On the one hand, the constructivist knowledge process can be regarded to be less *naïve* than the realistic one, because it requires a progressive detachment of the human subject from the object of observation; however, this also implies the unavoidable price of alienation (Hinchliffe 2007) through the production of discursive practices that turn matter into speeches, and speeches into (the only) reality. As a result, the social construction of nature "denaturalizes" nature and turns environmental issues into a dispute of concurrent discourses that contextualize our relation with nature in different moral, political, and aesthetic terms (Eder 1996b). The objectivity of scientific discourses itself is a cultural effect produced by the common agreement on what counts as good explanation, i.e., an agreement on what counts as confirming truth (Rorty 1979). This understanding of truth as social agreement (including agreement among the scientific community) reduces the authority of natural science in the environmental debate but upgrades discourses that are commonly regarded as nonscientific (ethical, political, social, etc.) to the rank of relevant positions to be considered in the public arena.

Constructivism suggests that nature is defined and delimited by different societies according to their thought categories, perceptions, and understandings. In this context, discourses turn to be the most relevant form of political activity, and nonhuman subjects and nondiscursive practices are pushed into the realm of mere objects of speech. Do forests exist? Yes, as long as they enter into discursive practices. Do hurricanes happen? Yes, insofar as they are included in political analysis. The way in which description is given *counts* as reality and the discursive mediation produces different social decisions (Adger et al. 2001). Obviously by taking the constructivist approach to its logical extreme, different constructions *actually* imply the existence of different worlds (Braun and Wainwright 2001). A radical constructivist approach puts into question the foundational assumption that nature in itself can be regarded as ground for social or ecological values (Castree 2001, 17): "[h]ow, for example, can eco-centrists claim that killing whales or destroying the Amazon is wrong, if we can no longer 'appeal to nature as a stable external source of nonhuman values'?" As a consequence of the refusal to consider external data relevant, constructivism often leads to suggesting that what counts as reality depends on the analyst's perspective and thus, there is no easy way to separate objective observation from social biases. Truth is not self-evident, but it is the effect of discursive practices of signification: "when social practices are contested in defense of 'nature' the nature defended is also already a cultural artifact, since what counts as nature emerges in and through discursive practices" (Braun and Wainwright 2001, 46).

Box 2.2   Campaigning for the rainforest. The Western construction of indigenous environmental culture
In their work on the construction of nature, geographer Bruce Brown and John Wainwright note that "When social practices are contested in defense of 'nature' the nature defended is also already a cultural artifact, since what counts as nature emerges in and trough discursive practices" (Braun and Wainwright 2001, 46). This observation is not trivial, as it clearly points out one of the distinctive traits of the constructivist position, and at the same time its principal fallacy. In fact, when the objective nature of realism turns into a discursively produced object, it enters the realm of constructivist

interpretation and, while becoming available for public scrutiny, it also loses most of its authority of uncontestable fact.

One of the clearest examples of social construction of nature is the social construction of the idea of "pristine nature" and the concern for its preservation that materializes in campaigns aimed at safeguarding tropical forests and indigenous people. In fact, environmental organizations' campaigns and international organization (e.g., the *World Council of Indigenous Peoples*) portrayed indigenous people as disinterested stewards of nature, whose richness and balance have been preserved up to now thanks to their ancestors' vernacular knowledge. Since the 1980s onward, in the postcolonial wave, the struggle over the protection of the last remnants of Eden has been associated with the struggle over indigenous people's identity and rights (Braun and Wainwright 2001) and became a common matter of concern for environmental thinking (with the *Rainforest Alliance, Accion Ecologica*, and *Survival International* being good examples of this). Wolfgang Sachs from the Wuppertal Institute pointed out the fallacy of realism, arguing that it

> construct[s] a reality that contains mountains of data, but no people. The data do not explain why Tuaregs are drive to exhaust water holes, or what makes Germans so obsessed with high speed on freeways; they do not point out who owns the timber shipped from the Amazon or which industry flourishes because of a polluted Mediterranean sea; and they are mute about the significance of forest trees for Indian tribals... In short, they provide a knowledge which is faceless and placeless... It offers data, but no context (Sachs 1993, 22).

With the aim of providing this context, constructivists, while debunking the realist primacy of objective data, took this a step further and reinvented nature itself. As a consequence, the process of nature interpretation turned into a process of nature *creation*; this determines, for instance, the paradox that a forest is regarded as authentic and pristine only if it located where it is supposed to be and clearly demarcated. On the one hand, this reveals nature to be, for a large part, a cultural fact; on the other, it superimposes meanings upon simple matters of fact. The case of Penan people in the Malaysian state of Sarawak, Borneo island, reported by the anthropologist Peter Brosius clarifies the point (Brosius 1997). The matter started in the 1980s when timber companies moved into highland inhabited by the hunter-gatherer Penan, and they spontaneously began an active resistance, supported by national and international environmental organizations, to assert their land rights and preserve the rainforest (Brosius 1997; Manser 1996). Environmental activists substantially contributed to the construction the images of the Penan and their landscape and deployed them in international campaigns; these representations had a persuasive strength, so that Penan assumed them as a real description of their own culture. They are presented as traditionally using their environment in a long-term sustainable way, spontaneously concerned with ecosystem's balance and believers in the holiness of plants because they are said to possess a soul and to be born from the same Earth that gave birth to people. By linking a specific culture to general meta-discourses about indigenous people, environmentalists are able to produce a narrative that appeals to preexisting categories, imbued with ethical values (endangered nature, sacred land, inviolable people, etc.) and confers upon indigenous people the authority of the guardians of the Earth's secrets.

This culturalist gaze, even when endorsing praiseworthy intents, produces stereotyped representations of indigenous people; as Brosius notes, this "makes generic precisely the diversity that it is trying to advance [by] imposing a falsely universalized quality on a range of peoples, and thereby collapsing precisely the diversity that defines them." (Brosius 1997, 64-5). A brief anecdote reported by Brosius exemplifies this. When field-working in the Sarawak, he was once walking with one of the Penan who pointed out to him a Belaβan tree, telling that medical essences can be extracted from it. The anthropologist immediately took note, as this is exactly the kind of knowledge he expected indigenous culture to be able to provide. But when he further inquired how the Penan people use this essence, he was answered that actually they have no idea; it was one of the environmentalists who was campaigning to protect the forest who told the Penan about the Belaβan trees.

The different realist and constructivist positions also bring about two opposing opinions of what is the best road toward sustainability. They advance the theories of ecological modernization and ecological *reflexive* modernization, respectively.

Realism proposes that the blueprint for a new ecological modernity can be led by economic competition and technological innovation, which produces, in turn, economic growth while using less energy and resources and producing less waste. This "win–win" strategy is generally warmly embraced (albeit not systematically implemented) by national governments and international institutions and requires the incremental adoption of market-based instruments in environmental policy. As philosopher Peter Hay (2002, 230) explains, it is particularly welcomed by business sectors too:

> Ecological modernization [...] assigns a central role to the invention, innovation and diffusion of new technologies and techniques. [...] Ecological modernisation constitutes an alternative, welcomed by governments with relief, to the crudely punitive regulatory regimes that have failed to deliver environmental protection or improvement since the 1970s.

A noticeable point is obviously that environmental NGOs, while often adopting a realist position in their campaigns, are in general more inclined to support a long-term strategy closer to the reflexive modernization approach (although some of them are happy to take part in business-oriented networks aimed at green development, e.g., the WWF's project "Changing the nature of business"). They correctly point out that improving the environmental efficiency of production processes through technological innovation is praiseworthy, but it does not reduce aggregate levels of resource consumption and waste production—on the contrary, gains in environmental efficiency typically fuel further consumption and production. Moreover, not all environmental protection measures (e.g., biodiversity protection or forest protection policies) are necessarily conducive to economic growth, and in some cases political trade-offs may be necessary if they are to be implemented. Finally, technologically driven ecological modernization offers very few possibilities to address the unequal distribution of ecological risks among different social classes and nations, nor means for correcting environmental injustices (Hay 2002).

These considerations make NGOs' positions closer to the ecological reflexive modernization approach, which has been proposed in response to the neoliberal mainstreaming of environmental thinking by a large number of reputed constructivist scholars. Among them, the sociologist Ulrich Beck explains that ecological problems persist because they are generated by the very economic, scientific, and political institutions that are called upon to solve them. Simply producing more environmentally efficient tools, therefore, cannot solve the paradox of sustainable development; rather, it is necessary to adopt a reflexive attitude toward the means *and* the ends of late modernization and its consequences. While theories of environmental modernity suggest that social and political protest directly derives from the objective and scientific analysis of urgent environmental problems, Beck opposes this view by claiming that the emergence of ecological issues is a matter of culture, not of scientific data:

> [T]he dying forests do not contain in themselves the reason for the public attention and concern they receive. Every attempt to deduce social and political protest from an objective, natural-scientific analysis of its urgency [...] derives from the prevailing technicist confusion between nature and society. (Beck 1988, 47)

As a result of the large-scale environmental hazards that the modern world has experienced, uncontested authority has been bestowed upon the technical language and practices of laboratories, but technology cannot guarantee safety unless it is combined with input and feedback from social institutions. This does not mean that ecological reflexive modernization theory does not require restriction, taxes, or controls (as ecological modernization does); rather, it means that these initiatives need to be complemented with a critique of the functioning of capitalist societies. Ecological modernization is not an ideology-free theory, for it is able to naturalize green capitalism and keep it safe from deeper questions and social unrest. The constructivism of Critical Thinkers thus supports the development of a more adequate cultural basis for environmental thinking and politics (Eder 1996b)[3] in order to gain widespread consensus on key environmental values, including social values (such as justice, equality, and participation).[4] The ecological reflexive modernization theory is, however, not itself exempt from critique, as it is essentially an attempt to rescue late modernity from the (supposed) deviations of post-modernism and to initiate a new Enlightenment by re-reading the unfulfilled project of modernity against the blackness of industrial society (Beck 1995). Although the constructivist approach is apparently more democratic than the realist one, it is based upon the assumption that there are universally shared green values and a consensus on the need to deal with environmental issues. Where people will not voluntarily endorse the changes sustainability requires, discursive democracy is offered as a solution so that they will see the advantages of ecological reflexive modernization and consequently embrace it. The stress on the discursive side (characteristic of the Critical Thinkers and, particularly, influenced by the seminal works of Jürgen Habermas), however, does not guarantee either consensus or universality and actually prevents those actors unable to take part in discourses from joining the cause.

2.2   Normalization and Mainstreaming

Despite the internal opposition between realist and constructivist approaches the increasing relevance of environmental thinking in the 1980s and 1990s went together with a process of progressive normalization and mainstreaming. The existence of different strategies for achieving sustainability goals (including the straight and strong positions of radical environmentalism) was not an impediment to the creation of an eco-liberal orthodoxy that granted experts (whether natural or social scientists, consultants or planners) the authority and legitimacy to elaborate "a set of principles and intentions used to guide decision making about human management of environmental capital and environmental services" (Roberts 2010, 2–3). This more palatable version of the radical environmentalist message (decrease consumption to decrease our impact on the environment) allowed international institutions and governments to pursue a process of progressive de-politicization of environmental thinking by removing claims that were potentially problematic in terms of the global balance of power, such as those more directly advocating for equality, redistribution of resources, and consumption choices, or condemning crimes against nature and people, as well as overexploitation of resources. In so doing, they were able to achieve a broad consensus on environmental values and gain public support for environmental policies. At the same time, however, the revolutionary message of environmental thinking was refashioned to become part of the international political mainstream as something everybody could be committed with—independently from other political, cultural, or social preferences. By stressing technical tools and procedures for achieving the agreed-upon targets of pollution reduction, green energy production, recycling, and similar, the groundbreaking message and the sociopolitical impact of environmental thinking got lost—literally—in the course of business as usual.

The mainstreaming of environmental thinking, which resulted in environmental values entering the lifes of ordinary people as well as being added to the international and local political agenda and companies' business plans, required it to undergo a preliminary process of normalization. This includes both an epistemological normalization, i. e., the identification of reliable discourses showing the truth about the state of the world and a political normalization that establishes the rationale for an agreed-upon normative social behavior (Foucault 2003). In his *Philosophy and the Mirror of Nature* (1980), Richard Rorty describes the origin and functioning of the epistemological normalization process. He explains that in the Western rationalist tradition, truth is actually not regarded as the property of correspondence between our assumptions and the external world, but rather as an agreement about what *counts as* accurate representation of it. Truth is, thus, under the authority of consensus, rather than of empirical observation. This means that producing an accurate representation is a matter of convergent opinions about statements whose content has proved to be useful in achieving common aims, and, as a consequence true is "simply an automatic and empty compliment which we pay to those beliefs which are successful in helping us do what we want to do" (Rorty 1980, 10). Normal discourses are those mirroring reality in the sense of producing a consensual truth and represent:

> [A]n agreed-upon set of conventions about what counts as a relevant contribution, what counts as answering a question, what counts as having a good argument for that answer or a good criticism of it. (Rorty 1980, 230)

The problem of accurate representation is of fundamental importance in environmental thinking, as the very legitimation of environmental claims and policy initiatives is based on the ability to demonstrate the cogency of specific issues. The development of environmental thinking is in fact largely dependent on the possibility of defining a stable epistemological foundation for the truth-seeking process that can resist the attacks of opposing claims and make different positions comparable.

The epistemological normalization process in environmental thinking does not have a value per se as a knowledge practice, but rather it is tightly connected with the political normalization process. This requires the definition of a corpus of axioms with normative power that are able to constraint and address individual and collective agency. In Michael Foucault's interpretation (Foucault 1977), they materialize into judgment, norms, indications, rules, and in general any pronouncements on what is a normal or abnormal behavior (which he argues are the basis for modern disciplinary institutions, such as schools, prisons, psychiatric hospitals, etc.) from the individual level of everyday micro-politics to the macro level of government. This translates into the definition and diffusion of social models that soon become sources of control over people's conduct.

While epistemological normalization is aimed at establishing what arguments count as good arguments in the definition of environmental truth, political normalization is aimed at establishing what arguments are good arguments in the definition of the socio-environmental order. Good arguments are based on knowledge. Knowledge, Foucault explains, is in a reciprocal relation with techniques for maintaining and enacting power, so that they may turn into effective instrument of government over people's decision and behaviors. In order for this to happen, the political normalization process requires that governments appreciate the effects on people of diverse conduct, education and management models, tactics of persuasion, incitement, and motivation. At the same time, normalization demarcated the border between the meaningful discourses, mainly those advanced by the eco-liberal orthodoxy, and the nonsensical claims of radical environmentalism. It translated the critical and highly politicized language of environmental movements into the more widely understandable language of environmental politics. The involvement of private investors in environmental projects is clearly indicative of the power of normalization processes, which have been able to turn anti-capitalist environmental values into something worth investing in.[5]

The political normalization of environmental thinking entailed, for instance, the elaboration of a liberal model of green citizenship (Dobson 2003b) whose environmental virtues include people's participation in collective decisions about their surrounding environment, adopting eco-friendly behaviors (e.g., recycling, choosing local food, avoiding plastic bags, and buying green) and supporting environmentally conscious political plans. All of them become part of normal life for a large portion of global citizens.

After the 1992 UN Summit on Environment and Development green-liberal discourses became part of normal life in most Western countries, common sense ideas that people were used to hearing on TV and reading about in newspapers. Geographer Noel Castree notes, however, that while normalization processes made it possible to mainstream environmental thinking, at the same time they generate three paradoxes that largely contributed to the discredit of environmentalism. First, environmental issues have been on the agenda of ruling parties in Western world for many years even if environmentalists refused to recognize it and acted as if their claims still were undervalued; second, environmental discourses have been appropriated by actors (such as institutions, private companies, and organizations) with very few green credentials, so that they remained nice-sounding statements with no effectuality;[6] finally, the public generally declared that it cares about the environment but did not necessarily translate this care into concrete actions (Castree 2006).

Box 2.3   The (ineffective) mainstreaming of the fight against climate change

The debate on climate change clearly exemplifies how mainstreaming environmental issues does not necessarily bring about effective results in terms of achieving sustainability targets. The issue of climate change became relevant for the wider public in the 1980s after a broad scientific consensus was reached among scientists on the anthropogenic causes of the problem (or, in Rorty's terms, the truth about the nature and reality of climate change was established among those whose opinion counts as authoritative enough to confirm it).

The first international convention on climate change was adopted during the Rio "Earth Summit" in 1992 (e.g., the UN Framework Convention on Climate Change, UNFCCC), which introduced for the first time the principle of common but differentiated responsibility. As it was a framework convention, its only real achievement was the recognition that industrialized countries are the main parties responsible for greenhouse gas emissions and thus are also principally responsible for reducing them and for delivering financial and technical assistance to developing countries in undertaking a different path. However, the UNFCCC did not include any legally binding emission reduction target; instead, it postponed the need for defining more precise commitments. It was thus the 1997 Kyoto conference that, after years of negotiations, introduced in the Kyoto Protocol some legally binding measures for developed countries committing to reduce their emissions by 5% compared to the 1990 level in a maximum of 15 years (Henson 2011). The agreed target was extremely modest compared to the recognized need to ultimately reduce emissions by 60–80% in order to limit global warming, but it was largely a compromise between the European Union, which proposed a 15% reduction for industrial countries, and the United States' desire to keep any emission reductions as small as possible. Moreover, the Kyoto Protocol had the important effect of adding the climate change issue to the list of top global priorities. It is also a particularly clear example of normalization of the potentially disrupting consequences of a serious inversion in socioeconomic development strategies by adopting technical regulatory norms and market-based tools. The Kyoto Protocol in fact established three mechanisms to allow countries to meet their obligations: the "joint implementation" (offering signing parties a flexible and cost-efficient means of fulfilling their commitments by earning emission reduction units from an emission-reduction or emission-elimination project in another country); the "clear development mechanism" (outlining the possibility to implement an emission-reduction project in developing countries and thus earn saleable certified emission reduction credits); and the "emissions trading scheme" (allowing countries that have emission units to spare to sell this excess capacity to countries that are over their targets). Together with the skepticism about the real capacity for the Kyoto mechanisms to produce actual changes instead of simply establishing a market for carbon (the principal greenhouse gas) just like any other commodity, the Protocol was also criticized by environmental thinkers and environmentalists for leaving unanswered issues of compliance and implementation, together with the more crucial issue of adaptation to climate change (Victor 2001).

These criticisms, together with strong disagreement about complementary contentious on geopolitical issues led the US president George Bush to withdraw from the Protocol in 2001, thus putting at risk—due to the United States' weight in terms of carbon emissions—the future of the agreement. Nonetheless, in 2005 with the entry of other major polluting countries such as Canada and Russia, the Protocol entered into force; disregarding the (quite modest) achievement of committing countries, this step conferred the climate change issue a key role in local and regional policies by entering into ordinary language and mind-set of ordinary people nearly everywhere in the world. From a green diplomacy perspective, the multiple incongruences and the weaknesses of the Protocol emerged clearly during the 2007 UN Climate Change Conference in Bali where the rising level of developing countries' emissions was brought to the fore by developed countries, who urged emerging economies to take up responsibilities for reducing emissions. At the same time, the developed countries themselves were found to be not doing enough to reduce their emissions; the European Union, in which until then had promoted efforts to reduce emissions in international negotiations, became reluctant to participate in the agreement. The negotiations during the Copenhagen Climate Change Conference in 2009 were disappointing, with no better results achieved during the Cancun Climate Change Conference 2010 or the Durban Climate Change Conference in 2011. The latter was only able to deliver an accord based on self-selected mitigation commitments, international monitoring, and reporting procedures.

However, from the perspective of the public imaginary, the mainstreaming of the climate change issue began to gather momentum when in 2007 the Nobel Prize for Peace was awarded to the Intergovernmental Panel on Climate Change (IPCC) (jointly with Al Gore) "for their efforts to build up and disseminate greater knowledge about man-made climate change, and to lay the foundations for the measures that are needed to counteract such change" (The Nobel Peace Prize 2007). The social and environmental consequences of global warming (including, for instance, the tragedy of environmental refugees (Collectif Argos 2010) become part of the public debate, with regional, local, and grassroots initiatives flagging cities as places for experimenting with innovative solutions to mitigate climate change. These range from the approach of the resilient cities movement (e.g., the campaign on resilient cities spearheaded by the ICLEI—Local Governments for Sustainability network) to the most radical approach of the Transition Town Movement (see this book, Chap. 4). All are inspired by the belief that bottom-up and local initiatives can succeed where top-down efforts have failed by bringing about a change in people's mind and behavior. The future will show to what extent this is actually possible.

In practice, environmental thinking became mainstream when it entered natural and social scientific research and gained institutional recognition; environmental issues have been largely popularized by the media, included in business strategies, and had an influence on humanitarian interventions. Sustainable development became the pillar of many sociocultural, environmental, political, and economic programs, supported by communication campaigns aimed at raising the public standards of collective morality and personal responsibility (Katz 1998). Most notably, regulatory norms and market-based instruments (e.g., the economic and financial tools fueling the emissions market defined by the Copenhagen Climate Change Conference in 2009) have been broadly discussed by an emerging branch of neoliberal economy called environmental economics (Heal 2000) and largely debated by the opposing ecological economics (which calls for a more radical restructuring of the economic system and proposes a new bioeconomy; Daly and Farley 2004).

Nonetheless, despite substantial joint efforts by international diplomacy and national agencies to arrive at a consensual environmental politics, the results did not live up to expectations, and the target goals defined by so many declarations and international agreements were often disregarded (UN 2002).

The entire parabola of environmental thinking thus describes its transformation "from a knowledge of opposition to a knowledge of domination" (Sachs 1993, xv). Although it started as a protest movement calling for new public democratic virtues, as soon as environmental issues moved to the top of the international agenda, environmentalism lost its critical force. It limited itself to suggesting better managerial strategies to support the global empowerment of governments and corporations, as well as survival strategies that do not effectively challenge the current socioeconomic order. The preferences accorded to science- and technology-based solutions for addressing environmental challenges weakened the political, social, and cultural strength of environmental claims and reduced social agency. At the same time, the excessive stress on the discursive and cultural side of environmental issues transformed most of them into matters of quasi-academic debate that ordinary people have little access to; thus environmentalism excluded itself from the everyday lives of the people. Far from achieving a deeper level of understanding and a higher level of objectivity or even commitment, environmental thinking finds itself trapped in a fatal interplay between the will of knowledge and the exercise of power. This means that the inability of mainstream environmental thinking to cope with ecological challenges does not merely derive from *external* conditions, but rather from *internal* constraints determined by its very nature, objectives, and means.

NOTES

1. Amongst the proto-ecologists, the Danish botanist Eugenius Warming (1841–1924) and the US botanist Frederic Edward Clements (1874–1945) advanced the idea of nature stability, as did the early ecologist Arthur Tansley (1935).
2. See, for instance, the report by Nicolas Stern's review *The Economics of Climate Change*, available at the HM Treasury web page, http://www. hm-treasury.gov.uk/independent_reviews/stern_review_economics_cli mate_change/sternreview_index.cfm.
3. There are several definitions of political ecology proposed as alternatives to environmental politics or environmentalism; the supporters of political ecology usually interpreted it as able to subvert the apolitical conception of environmentalism. Eder adopted this last version and defined political ecology as *realism* in green politics.
4. Eder proposes the Deep Ecology movement as an example of a reaction to economic accumulation and obsession with growth. This movement increases the potential for self-expression and creativity. Deep Ecology (or Ecosophy) is a philosophy of the 1980s, based on a shift from the so-called anthropocentric bias of established environmental movements, which are censured for having a utilitarian and anthropocentric attitude toward nature. It is defined as *deep* because it asks complex and spiritual questions about the role of human life in the ecosphere, seeks to end authoritarianism through decentralization, and espouses a less dominating and aggressive posture toward nature. In fact, Deep Ecologists support decentralization and the creation of ecoregions, the breakdown of industrialism in its current form and the end of authoritarianism. Arne Naess, Bill Devall, and George Session are some of the principal inspirers.
5. This is the ecological alternative advanced by environmental economists, whose proposal for greening politics is mainly based on market solutions, including eco-taxes for polluters, emissions trading to control pollution, and economic incentives. Both the solutions are developed from neoclassical economic theory and are based on the belief that private corporations and public authorities, thanks to market mechanisms, would be able to manage environmental restoration.
6. In formulating this paradox, Castree is influenced by Klaus Eder's work (1996b).

CHAPTER 3

Is This the End of Environmentalism, as We Know It?

Abstract This chapter explores the theoretical critiques that environmental politics attracted since the 1990s onward. Among them, building upon political ecology critique, post-environmentalist theory gained a prominent role and claimed that green diplomacy, business, and large NGOs determined a de-politicization of environmental issues in the pursuit of establishing a widespread consensus on the mainstream strategies for global environmental governance. The origins and development of post-environmentalism are described in the chapter, with particular attention devoted to the differences between the realist perspective of US scholars (most notably Michel Shellenberger and Ted Nordhaus, authors of the pamphlet "The Death of Environmentalism") and the European scholars, advancing a constructivist interpretation of the end of environmentalism under the name of post-ecologism.

Keywords Political ecology · eco-cracy · Shellenberger and Nordhaus · Blühdorn

Since the 1990s, a number of theoretical critiques addressing the internal weaknesses of the sustainable development paradigm and the inefficacy of global environmental policies have emerged in the environmentalism debate. These critiques, born of many diverse voices that merged together both realist and constructivist interpretations, urged environmental thinking

© The Author(s) 2016
C. Certomà, *Postenvironmentalism*,
DOI 10.1057/978-1-137-50790-7_3

as a whole to take a step toward a new appreciation for the complex relationship between the environment and society. In fact, it is not accidental that many of the most vocal environmental critics have manifested as very public forms of political or social ecologists, presumably because within their common character resides strong interest not solely in the causes, but also in the social consequences of environmental issues. More specifically, those acting as political ecologists (Robbins 2004) spotlight the post-structuralist, neo-Marxist tradition that focuses on the links between environmental degradation and socioeconomic marginalization with the aim of contesting the increasingly apolitical understanding of ecological challenges (Forsyth 2008). This theory emerges at the confluence between ecological oriented studies in the social sciences and principles of the political economy and demonstrates how political power relationships play a crucial role in determining the unequal inter and intra-generational distribution of costs and benefits through society, in most cases by reinforcing already existing social and economic inequalities (Bryant and Bailey 1997).

3.1   THE POLITICAL ECOLOGY FRAMEWORK

Generally speaking, political ecology recognizes the importance of gathering reliable data for determining appropriate strategies to tackle environmental problems. It also acknowledges that some aspects of environmental problems are difficult to monitor, quantify, assess, or predict, as these measurements also pertain to the sphere of social, cultural, and political interactions. Different from previous environmental theories, it incudes to a large extent the field experience of environmentalists and other various social movements, which are now recognized as key actors in the global political sphere (Sassen 1998), and benefits from the production and diffusion of non-western contributions to the environmental scholarly debate. This makes it possible for environmental thinkers, including those in academia, to openly acknowledge the values of actors other than just those providing official science-based forms of knowledge, and the impact of globally dispersed communities of practices in the new geometries of power of the emerging geopolitical order (Castells 1997).

Thus a large and varied number of political ecology theories from the real life experiences of people from around the world found its place in the panorama of environmental theory discussion, including most notably the postcolonial environmentalism and the environmentalism of the poor (Alier 2003), environmental conflicts (Homer-Dixon 1999), and the environmental justice theories (Agyeman 2005). All of them share a common understanding of mainstream environmental thinking as a form of eco-cracy, that is, a powerful normative and universalist theory hidden under the aimed-for-better intentions of international environmental regimes, which calls for effective managerial strategies to support global empowerment of governments and corporations, and to rationalize lifestyle in order to provide survival strategies without effectively challenging the current socioeconomic order. Philosopher Val Plumwood was the first to introduce the term eco-cracy (Plumwood 2002), where she metaphorically described the structure and functioning of an imaginary eco-republic, or a global version of Plato's rationalist utopian republic led by hyper rational decision makers, able to precisely identify environmental emergences and to propose appropriate solutions thanks to the support of natural scientists. Plumwood explains that mainstream environmental thinkers actually seem to be interested in establishing such an eco-republic because this will lead people to live in a way which do not generate environmental problems, or which would generate problems of lesser significance than was previously the case. However, this implies that a preliminary agreement on what environmental concerns are to be addressed have been already reached on the basis of uncontested scientific evidence, which then provided the basis for a broad and ideally universal consensus. It also implies that a sort of Baconian benevolent despot will be granted with the power of governing the global eco-republic with the help of natural laws scholars and jurists.

While recognizing that such an eco-republic does not actually exist, Plumwood nevertheless warns against the temptation for global institutions to delegate political decisions to experts, in the name of risk prevention and environmental damages mitigation. In fact, Plumwood found that the eco-cratic approach of mainstream environmental thinking could potentially result in excluding from the public debate those who are most likely to suffer from environmental degradation, pollution, and hazards:

> [T]hose who have most access to political voice and decision–making power to be also those most relatively remote from the ecological degradation it fosters, and those who tend to be least remote from ecological degradation and who bear the worst ecological consequences to have the least access to voice and decision power (87).

These last considerations represent the crossing-point for the eco-cracy critique to meet political ecology. Without a constant and in-depth attention toward the mechanisms of environmental crises reproduction, no technically based political intervention may alleviate their consequences, often because these are generated and perpetuated by the same economic, scientific, and political institutions that are called upon to solve them.

Political ecologists owe much of their critical verve to post-structuralist analyses. Particularly, the postcolonial studies of Said (1977) and Pratt (1992) provided an invaluable source of inspiration to the western world in sharing the richness of non-western theoretical and practical elaborations. Postcolonial scholars criticize environmental politics as a further form of colonialism, hidden under the neutral appearance of scientific evidence and urged by the imminence of environmental risks where environmental politics are described as the paternalistic power of the western world aimed at imposing western values on the non-western world. Thereafter, so-called postcolonial environmentalism contests the normalizing power of mainstream environmental thinking, whose power consists of the subtle ability of establishing and imposing a single vision of truth supported by scientific results and popularized by international institutions. However, this apparently objective truth is often reputed as nonsensical in non-western cultures, or, more negatively is regarded as an attempt of impeding non-western people from establishing their own development model (Shiva 1993). Postcolonial environmentalists repute a serious consideration of non-western and nonmainstream cultures as the first and essential step toward the empowerment of local and indigenous people (Huggan and Tiffin 2010; Ross and Hunt 2010). Despite some environmentalist campaigns transformed the local and indigenous peoples in a romantic stereotype (that produced the most paradoxical effect of shifting from the colonial excess of denying local and indigenous people to the postcolonial excess of fetishizing them (Katz 1998)).[1] Postcolonial environmentalism has nevertheless given rise to one of the most important theories developed in the 1990 panorama of environmental thinking, i.e. the environmentalism of the poor. The environmentalism of the poor theory, as the economist Joan Martinez Alier presents it, transposes postcolonial instances in the political economy framework (2003). Martinez Alier explains that northern countries are increasingly dependent on the imports from the south, primarily of raw materials or consumption goods. More than ever before, this progressive erosion is pushing the boundaries of natural resources exploitation and is responsible for generating dramatic social and environmental unbalances, especially determined by the geographical displacement of sources and sinks and by their impact on ecosystems (Martinez Alier 2003). Despite the optimism of the ecological modernization paradigm, these unbalances cannot be readdressed by economic policy or changes in technology, as they are disproportionately borne by social groups who have no voice to denounce the uncontrolled environmental degradation and human oppression, more possess the political weight to reverse the current geographies of power. Thus, the environmentalism of the poor theory denounces the inevitability of ecological distribution conflicts determined by the fact that with the increase of economy

> more waste is produced, natural systems are damaged, the rights of future generations are undermined, knowledge of plant generic resources is lost, some groups of the present generation are deprived of access to environmental resources and services, and they endure a disproportionate amount of pollution. (Alier 2003, 12)

The ethical values motivating environmentalist resistance in the global south are profoundly different from those inspiring conservationists' concern for the loss of wilderness in the north. While the actors of the environmentalism of the poor are often peasants, local and indigenous peoples claiming to have coevolved with the environment and to have been responsible for maintaining biodiversity and conservation of natural resources for centuries, with the globalization of social movement struggles and the advent of internet and digital communication, these groups are progressively increasing their capability of confronting mainstream environmental thinking of northern countries (EJOLT 2015). For instance, a large movement supported by scholarly research argues that the economic debt claimed by northern countries has been already repaid by the ecological debt determined by centuries of overexploitation, violence, and rapes perpetuated by northern countries in the global south (Donoso Game 2009), and by the illicit use of traditional knowledge of seeds, ecology, and nature in general for private companies' profit, now generally defined as a form of "biopiracy" (Shiva 1993).

Despite being deeply rooted in post-colonialism, the environmentalism of the poor also takes inspiration from another revolutionary approach in political ecology originating in the migrant communities of North America, or the well-known environmental justice movement. Environmental justice theory points out that environmental injustices occur when unaccountable social agents externalize the environmental costs of their decisions and practices to innocent third parties in circumstances when the affected parties, or their representatives, have no knowledge of or input in the ecological risk-generating decisions and practices (Alier 2003). In the global context, it may also occur that environmental injustices occur when nations or global elites appropriate more than their fair share of the environment, however quantified, and leave behind oversized ecological footprints (Wackernagel and Rees 1998). Led by Chico Mendes in Brazil, the struggle for the preservation of the rainforest jointly with workers' rights exemplifies how the connection between environmentalism of the poor and environmental justice dates back to the 1970s (Gomercindo and Rabben 2007).

The environmental justice movement finds its inspiration in the spatial understanding of social justice scholars since the 1970s, working in the wave of the spatial turns of the social sciences. This helps to demonstrate that the circumstances in which different social groups live play a major role in determining their material wealth, opportunity, health outcomes, educational attainment, job creation, and virtually all of the metrics of quality of life which are never equally distributed across space (Harvey 1973; Lefebvre 1991; Soja 1989). The one-size-fits-all liberal model of social justice is an ideal with no possibility of real application (Harvey 1996) because the actual distribution of opportunities, material and nonmaterial benefits, services, and resources are not equally distributed through the space (Soja 2010). Moreover, this unequal distribution overlaps the unequal wealth and power distribution occurring through the social body (I.M. Young 1990). Several decades later, when the global population became aware of the pervasiveness of environmental degradation, pollution, and its dreadful impacts on social communities, the environmental justice theory acquired popularity (Dobson 1998). Empirical studies demonstrated that significant differences exist in the socioeconomic status of those communities living nearby environmental hazardous facilities or polluted areas, and those that do not (Mennis 2002; Faburel 2010). What makes environmental justice a sociopolitical theory is the fact that although ultimately no one can entirely avoid risk (Beck 1995), it does not mean that inequalities in the distribution of risk do not exist. Environmental problems are not randomly distributed; they do affect some people more than others. Although environmental justice is often adopted to address large-scale global issues, such as climate change or desertification (Agyeman 2005), it also manifests in the ordinary day-to-day life of local places where the consequences of environmental problems can reinforce already existing inequalities (Sachs 1993; Haughton 1999). Political philosopher John Dryzek (1987) notes that this condition is often determine by a "displacement problem," meaning environmental problems are apparently solved by moving them to another time (e.g. to future generations), to another space (e.g. another country), or even to another medium (e.g. by turning air pollution into water pollution). When an apparent environmental problem is displaced, one does not have to acknowledge its existence and therefore does not have to seek mitigation or resolution measures. Of course, marginalized groups and ethnic minorities are those most likely to suffer from pernicious environmental problems generated by displacement processes because they lack political and economic power, and this, in turn, gives them fewer opportunities to counteract poverty and social discrimination (Agyeman 2005). Here a further similarity emerges with the environmentalism of the poor theory, as this condition does not prevent marginalized and poor people from joining the global struggle against environmental injustice taking the form of environmental conflicts (Bromberg et al. 2007). Scarcity of resources and environmental deterioration exacerbate environmental injustices and determine the emergence of social conflicts clearly related to the ecological conditions and their economic and political consequences. Environmental conflicts are not only intended for reclaiming ecological restoration or damages prevention, but also for the empowerment of marginalized groups, an empowerment that can emerge as a result of a virtuous relationship between participation, recognition, and redistribution (Fraser 1997).

The multiple theories and practices emerging in the critical context of political ecology profoundly challenged mainstream environmental thinking, particularly by exploring the relationships between social and ecological issues (Leonard and Kedzior 2014). This approach also largely pervades the most recent post-environmentalist critiques.

3.2   Prophets of the End: The Emergence of Post-environmentalism

Scholars and activists advocating for the need of a post-environmentalist understanding of the current development of environmental thinking and practices did establish it as a single coherent discourse on the weaknesses of mainstream theory, and their contributes have been occasionally mentioned in larger historical analyses of the environmental movement as the edge of environmental debate (Haq and Paul 2011). In time, the term "post-environmentalism" has been appropriated by many kinds of environmental movements, including traditional conservation organizations (Neri 2004), or even bioregionalist exponents of the deep ecology (Berg 2001). Although diverse approaches advance different proposals on how to transform future environmental thinking, they share a common critical perspective on its mainstream interpretation. Specifically, while post-structuralist scholars address the weaknesses of the global environmental regime in the broader context of the critique of late modernity, post-environmentalists focus on the internal reasons that determine a crisis of reliability and effectiveness of environmental politics.

Though rarely mentioned, the birth of post-environmentalism dates back to 1990 when John Young introduced the term in the international scholarly debate by delivering an anonymous book, which linked ecological concerns to other political and social issues and analyzed the broader consequences of industrial culture. The intent of Young's book, though ambitious, was to show "why the problem of what to do about it [i.e. about the fact that we are living in a finite world and we have the power to destroy it] became not merely technical but also economic, political and moral" (1990, ix). The vastness of the intervention field described by Young can be fully understood when considering that together with political ecology, Young's work is influenced by postmodernism (Lyotard 1984). It offers clarification on multiple, disaggregated and partially overlapping positions where the general aim is to react to the generalized crisis of postmodern society, which in Young's account is characterized by "destructive, alienating technologies and a moral vacuum which has allowed the persistence of serious social inequality and an ill-fated exploitative relationship with the natural world" (Baker 1991, 204). Young noted that during the 1980s, environmental issues moved from the periphery of the political agenda to the center, so much so that green votes in many countries became a constituency for major ruling coalitions. With the awakening of environmental consciousness in the second half of the twentieth century, many described environmental problems as a consequence of late modernity (Giddens 1990; Taylor 1992), but at the same time succumbed to "the suggestion of scientific and technological determinism and believed that environmental problems were scientific in nature [and therefore] science could be expected to provide solutions" (Young 1990, ix). The situation raised itself to saturation level when conservative and liberal politicians co-opted green agenda so that they were able "to promise increased value-free economic growth in the context of a deregulated global market" (Young 1990, 117). Consequently, the green movement won a number of very important battles but lost the war against the consumerist ideology. In continuity with political ecology suggestions, Young pointed out that this mainstream approach produced a de-politicization of environmental issues, by removing from public sight more controversial side-implications, in the pursuit of establishing a large consensus on the mainstream strategies for establishing global environmental governance. The process went together with the search for technical solutions that disempowered social agency and produced largely ineffective policy measures. However, differently from much of the environmental thinking of the 1970s, Young did not speak out against technology or economic growth but recognized that halting growth would generate massive social conflicts. Rather, post-environmentalism provides the framework for different kinds of people undertaking a trial and error process toward the definition and realization of a sustainable society. All of them recognize that action is necessary but do not necessarily agree on the course of action to be undertaken and adopt different political and ethical behaviors. Young suggests the Gaia hypothesis and the eco-anarchist of Murray Bookchin (1983), or even Petr Kropotkin's anarchism, as powerful sources of inspirations for contrasting the ecologically destructive and unjust impacts of a capital-intensive development model. Moreover, this all-encompassing way of thinking can also be coupled with a wide range of technological choices, rather than tiny improvements of mechanical nature, that can better lead to a real fusion of science, politics, and ethics for a transition toward postindustrial society, as prefigured by the seminal Ernst Schumacher' work *Small is beautiful* (1973). Young auspicates a smooth reformist process operating through the long-term paths of democratic social transformation. This is the reason why Young's post-environmentalist theory has been regarded as a kind of pacificatory strategy with different actors proposing and adopting different ways for materializing their environmental sensitivity (Baker 1991). In line with the postmodern approach, it does not require engagement in terms of common ideologies, but rather in terms of common issues to be addressed.

Yet for some time Young's post-environmentalist theory remained an isolated attempt at sketching the future of environmentalism, and at overcoming the impasse of the progressive erasure of environmental thinking determined by the normalization efforts of international environmental

diplomacy and toning down the subversive claims of the early environmentalism hardline. So, despite Young's introduction of the word in
1990, post-environmentalism itself only become popular some years
later when it actively entered activists' discussions, and then later the
academic community.

*The Death of Environmentalism: Global Warming Politics in a Post-
Environmental World,* a rather long article was written by environmental consultants Michael Shellenberger and Ted Nordhaus in 2004
and published in part through *The Breakthrough Institute* and *Evans/
McDonough* research firm, gave rise to a large debate in the US public
policy sector by entering the debate on national institutionalized
environmentalism (Buell 2003). In particular, it contested the consequences of the mainstreaming of environmental thinking in public
policy from the first systematic measures adopted by the Clinton
administration, including the Kyoto protocol and the significant bud-
get increases for the Environmental Protection Agency (EPA), up to
the renewed centrality of environmental concerns under the Obama
administration.

Box 3.1   A Green New Deal for the post-environmentalist age?
The strategy of the Obama administration for dealing with climate
change and global environmental issues in general is particularly significant when exploring the mainstreaming of environmental thinking.
In this context, the Green New Deal program (Collina and Poff 2009)
fostered the introduction of green energy production industry as the
main partners in the process of advancing a sustainable development
strategy in the United States.

Elaborated against the background of the positivist sustainable
development concept, this was intended as a double-sided survival
strategy both encompassing environmental protection and economic
growth on the belief that "it is necessary to reduce carbon dependency and ecological scarcity not just because of environmental
concerns but because this is the correct and only way to revitalize
the economy on a more sustained basis" (Barbier 2010, 5).
Paradoxically as it may seem when considering this from an environmental activist's perspective, these new policies are essentially based
on the belief that world economy will be rescued by turning green.

The Global Green New Deal (GGND) (2008) strategy was launched by United Nations Environment Programme (UNEP) when the financial crisis transformed into a crisis or real economy and called for an urgent analysis of the role of green values in the economic recovery. The kick-off report was commissioned by UNEP to US economist Edward Barbier who produced a blueprint for the last offspring of neoliberal environmental economics, or the "green economy," published under the title *A Global Green New Deal: Rethinking Economic Recovery (2009)*. The title evocatively recalls former US President Franklin Roosevelt's New Deal strategy, which recovered the US economy after the financial crisis of the 1930s, and thus proposes green economy measures to enable the same recovery after the financial collapse of 2008. And while similar in platform, the Green New Deal does not ignore the macroscopic differences between the two historical contexts and clearly recognizes that development strategies adopted at the beginning of the twentieth century and now one century later need to strikingly differentiate on the basis in the difference of attention paid to the environmental costs of economic growth. Several countries across the world subscribed to the GGND objectives of reviving "the world economy, [creating] employment opportunities and [protecting] vulnerable groups; [reducing] carbon dependency, ecosystem degradation and water scarcity; [furthering] the Millennium Development Goal of ending extreme world poverty by 2015" (Barbier 2010, 8). The European platform for the implementation of the GGND, for instance, was intended to create the stimulus for a widespread change in people's lifestyles to be achieved by

> re-regulating the financial industry and channeling huge amounts of money into green investment to fund renewable energies, energy efficiency, sustainable industries and infrastructure, sustainable mobility, resource efficiency, protect natural resources and related research, innovation, education and training [to] stimulate entrepreneurship and create jobs. (Green New Deal 2014)

All of these initiatives needed to be complemented with new environmental standards, including the strict adherence to the agreed 20%

renewable energy share by 2020 and 45% by 2030 as a minimum, binding targets of at least 30% effective reduction in domestic greenhouse gas emissions in the EU by 2020 and a 60% reduction by 2030 compared to 1990 levels, environmental taxation in line with the "polluter-pays" principle, including a carbon tax, and financial incentives to virtuous initiatives (Green New Deal 2014).

The GGND was particularly popular in the United States where in 2009 green economy was strongly supported by US President Barack Obama, for tackling with both environmental and economic crisis. In his first economic discourses, President Obama urged to pass legislation that, with the aim of addressing the twin challenges of an ailing economy and the threat of global warming, committed the United States to double alternative energy production over the following 3 years and to the creation of a new electricity smart grid, while also promising to modernize 75% of federal buildings and to improve energy efficiency in over 2 million homes (Melvin 2009). Technical, financial, and regulatory innovation was thus to be adopted as the main pillar for sustainability, and President Obama was confident that the increase in the demand for solar energy equipment and alternative energy supply, as well as the degree of development and production required to produce them, would create new jobs and carry the burden of supporting a shift in economic growth from hard-manufacturing to alternative energy production. And with the intent of guiding, the United States toward a future less reliant on imported oil also came the realization of new offshore oil drilling and nuclear power generation opportunities, which most environmentalists would not support, but opportunistically fit preexisting business' way to sustainability. This also resulted in new financial and economic resolutions for the "creation of a trillion-dollar market in carbon pollution credits, billions of dollars of new government spending on breakthrough technologies and a tolerance for higher energy prices by consumers and businesses" (Broder 2011). Yet in the years following, the efforts of the GGND program were not as advantageous as expected. Examples of success reported by the European platform are rather modest and principally include small-scale initiatives inspired by autonomous green energy production, local initiatives of green consumerism, and best practices in public mobility here and there. In the advent of major international events such as the Deepwater Horizon or the British Petroleum oil spill of 2010, the 2014 earthquake in Japan leading to release of radioactivity at the Fukushima Daiichi reactor complex, and further ongoing decline of the US economic crisis, even the US strategy faded among the greenest ambitions and was only able to deliver some shallow measures of oil dependency reduction and clean electricity generation.

The green movement in US politics, though more boasted than implemented especially because of the internal congressional opposition, persuaded many environmentalist groups such as the *Sierra Club* or the *Nature Conservancy* that it would have been profitable for the advancement of their own cause to cooperate with institutions, but in so doing, however, their message lost much visionary and inspiring character, and their business turned into trivial negotiations and compromise seeking. Moreover, while legislative measures for the protection of air, water, and endangered species were effective in previous decades, similar successes are difficult to be replicated when global challenges, such as climate change, require a joint decisive effort. Consequently, environmental concerns, so staunchly included in the governmental agenda, actually started to lose their appeal, and the rhetoric of failure replaced the rhetoric of green development.

The Death of Environmentalism is articulated in two sections. In the first and longer section the authors argue that, after successful environmental campaigns which resulted in the development and implementation of a number of environmental laws in North America during the 1960s and 1970s, current environmentalism regards ecological issues as a sectional interest and not as the core of political agenda. This consideration induces a sense of shallow enthusiasm for the protection of a thing called the "environment," but no mobilization toward the real fulfillment of an alternative worldview. This literal interpretation of the objective, the purpose, and the identity of environmentalists transforms them into a special interest group and debilitates popular understanding because it advances a narrow definition of the group's primary interest of concern as merely focusing on the environment and excluding social issues. Schellenberger and Nordhaus conducted corporate marketing research to gather opinions, values, and beliefs of environmental communities across the United States and Canada. They built their argument upon a number of interviews with environmental thinkers, community leaders, and program funders in order to investigate the reason why modern environmentalism is increasingly unable to deal with the most pressing of ecological challenges notwithstanding the massive economic investment in them. From the collected interviews, the authors suggest that the responsibility for the loss of the environmental movement's authority is in actual fact due to the environmentalists themselves. They affirm to be on the right track but despite the efforts, current progress does not meet planned projection.

Most of the major environmental groups in the United States adopt a three-stage strategic framework, including the definition of a problem as environmental, the provision of technical remedy, and the selling of this remedy to legislators by using a variety of political tactics, such as lobbying, public relations, advertising, and so forth. When applied to global warming, this strategy led to unsuccessful initiatives in the global arena, and their failure was a direct consequence of the environmental movement's reductive view about the deep causes of global warming because "the environmental community had still not come up with an inspiring vision, much less a legislative proposal, that a majority of Americans could get excited about" (Shellenberger and Nordhaus 2004, 16).

Shellenberger and Nordhaus recount the story of fuel efficiency regulation in the United States as an example of environmentalist strategy to contrast carbon emission. They state that on several occasions since the 1970s, the environmental movement missed the opportunity to form effective alliances with industry and unions by adopting a narrow view of what the environmental struggle was about. Resultantly, environmentalists not only failed to win a legislative agreement on carbon, but also to determine the fuel efficiency standard, to impact US government policies for automobile technologies innovation, and to regulate the US automotive industry.

The reasons for the environmental movement's failure reside in the way the movement itself categorizes certain problems as environmental and other as non-environmental. In fact, it disregards the latter altogether with not having crafting any comprehensive political strategy to form larger alliances with other groups, with exception made only when they are of direct use for environmentalists' causes. Schellenberg and Nordhaus (2004, 12) contemplate:

Why, for instance, is a human-made phenomenon like global warming – which may kill hundreds of millions of human beings over the next century – considered 'environmental'? Why are poverty and war not considered environmental problems while global warming is? What are the implications of framing global warming as an environment problem – and handing off the responsibility for dealing with it to 'environmentalists'?

These questions closely recall the political ecology critique to the reductive presentation of the causes of environmental issues.[2]

In the second and shorter part of their pamphlet, the authors affirm that environmentalists have been too timid in raising the alarm about global warming, and that their focus on technical solutions lead to the adoption of very short-term policies with no remarkable consequences. The excessive attention for dedicated policies downplays the relevance of the politics that support them, but as long as the failures of environmental movement will be understood as essentially tactical and the solutions essentially technical, it will not be possible to affect any change. An exclusive interest for technical policies, like pollution controls, weakens environmentalism's popular inspiration without incrementing its political power. State Shellenberger and Nordhaus (2004, 8), "[w]e believe that the environmental movement's foundational concepts, its method for framing legislative proposals, and its very institutions are outmoded." The progressive transformation of grassroots environmental ideology into a set of tools advancing a normative, universalist, and authoritarian view of environmental governance, in fact, dooms to failure the efforts of many environmental organizations. Consequently, the increasing attention toward environmental issues does not correspond today to a strengthening of environmental thinking. Shellenberger and Nordhaus state that ecological issues are taken as sectional interests and thus merely induce a sense of shallow enthusiasm but no real mobilization, because when mainstream environmentalism advocates for green economic growth this paradoxically works as an obstacle for the realization of ecological values themselves (Buck 2012).

A number of critiques are thus advanced by post-environmentalists toward traditional environmental thinking, including

its perpetuation of a 'limits to growth' narrative which seeks to constrain human intrusions into the environment, its reliance on scare tactics to draw attention to the climate crisis, its emphasis on the need for people to make personal sacrifices, and its insistence on framing climate change first and foremost as an environmental issue at a time when most people are primarily concerned with economic security (Buck 2012, 2).

But Shellenberger and Nordhaus also invite one to consider the possibility for the elaboration of an alternative strategy based on the re-narration of the environmental movement. Environmental crisis is thus portrayed as an opportunity to promote ecological values while advancing economic prosperity and growth, on the belief that material wealth is most important for the emergence of environmental post-material values (Nordhaus and Shellenberger 2010). They claim that new bridge values framed around the core American values can lead to a more coherent and collective project able to attract wider consensus and greater investments, to create new jobs, to free the United States from oil dependence, to improve investments into clean energy, and to offer an inspiring vision of civil and unions rights, of business, and the environment (Nordhaus and Shellenberger 2004).

3.3   Post-ecologism. A Constructivist Interpretation of Post-environmentalism

While US-based critiques mainly focus on the technicalization of mainstream environmentalism, the European debate focuses on the possibility to reconsider its sociocultural and eventually ethical contributions (Hayward 1994). By building upon the late critical scholars' interpretation of environmental crisis as a crisis of modernity itself, the European understanding of post-environmentalism adopts a constructivist approach and suggests that environmental issues are social constructions, and that it is the very understanding of nature and its social representation that need to be questioned in order to overcome the current crises of environmental thinking (Eder 1996a).

The word post-environmentalism has been primarily adopted by critical thinker Klaus Eder, who elaborated a theory of practical rationality explaining that the current limits of the "ecological reason," or rather the reasons motivating environmental agency, are determined by the definition of nature as an object of human needs (Eder 1996b). Eder's intent is to contribute to a progressive resolution of symbolic conflict between the society and nature by introducing the idea of socialized nature. This is of fundamental importance for reaffirming the political spirit of environmentalism and can eventually lead to a new ecological reflexive modernity:

[T]he age of environmentalism, the collective mobilisation for a cause, is over. The age of post-environmentalism begins when ecology is established as a masterframe that can be referred to by all actors, thus laying the ground for a further development of the public space which is genuine modern condition for guaranteeing the cognitive, moral and aesthetic rationality inherent in the culture of modernity (Eder 1996b, 216).

Post-environmentalism, in Eder's view, can thus be the master frame for the development of the cognitive and moral modern rationality, which is characterized by the establishment of environmental issues as noncontroversial collective concerns (Eder 1996b). The consensual definition of environmental priorities calls for experts' evaluations of competing claims that can transcend the classic left versus right political dispute and command widespread agreement on environmental values. Inspired by Johan Rawls's theory of justice (1971) and Jurgen Habermas' theory of communication (1984), Eder proposes an understanding of post-environmentalism as a substantial set of principles and prescriptions based on the ideals of justice, equality, and participation (Shiller 2005). This of course means that rather than qualifying itself as universalist discourse aimed at greening modernity, such as Schellenberger and Nordhaus propose, the post-environmentalist discourse in Eder's view is intended to support a turn toward the post-industrial paradigm:

[T]he transformation of environmentalism into ecological politics is the central mechanism by which modern society learns to overcome the limits of the cultural model of early modernity and to develop more adequate cultural grounds for a democratic polity (Eder 1996b, 6).

The passage from environmentalism to ecological politics (which for the sake of this book can be read as environmental politics) is supported by Critical Thinkers as a crucial evolution toward a well-organized ideology able to "survive the market place of public discourse on the environment" (Eder 1996b, 165). Eder claims that we are now experiencing the integration of established environmental values in political process. Such integration, in line with the Habermas' work (1984), is understood as essentially the result of discursive communication, such discourses being the essence of political life. Moreover, as Habermas himself is situationist for what concerns the content of the discursive agreement (i.e. what is right, and good is to be defined by the participants in the debate), also Eder does not propose any substantive content for post-environmental values nor any specific technical solutions for environmental problems. In order to find an agreement between ideals and reality, environmentalists ultimately need to rely on experts' opinion. This means that, despite pledging the re-politicization of environmental issues, as realists do, also constructivists end up with aspiring to an environmental politics purified from conflict, ambiguity, and uncertainty. Featherstone (2002, 28) describes this as "an 'avowedly apolitical' approach to sustainable development structured by a 'rhetoric of partnership and stakeholder democracy' and a desire to achieve the global consensus amongst both citizens and governments'."

By advancing the internal debate in the critical scholarly community, mainly through dialogue with sociologist Ulrich Beck, philosopher Ingolfur Blühdorn (2000) worked extensively on the critical analysis of post-environmentalism, and identified the current form of environmental thinking as "post-ecologism."[3] Despite the fact that the debate itself had no broad resonance outside of the academia, it is nonetheless worthwhile of attention here as it exemplary represents the constructivist reaction to the post-environmentalist proposal. Blühdorn notes that neither Young nor Eder, Schellenberger and Nordhaus envisage a theory that goes beyond the ontological and epistemological foundations of environmentalism itself (Blühdorn 2000). In order to fill this gap, he attempts to examine what Beck called the "crisis of ecological crisis" by building upon the consideration that there is no possibility to progress any further by revolving around the external nature, but rather it is the human self and the identity construction that need to be rethought. In so doing, he shifts "the focus of attention from specific conditions in the physical environment which, according to ecologists, constitute objectively existing ecological problems, to certain cultural parameters which determine the way in which physical environmental conditions and their change are socially perceived and constructed as problematic or unproblematic" (Blühdorn 2000, 200).[4] This preliminary consideration suggests that constructivists need to engage in a critical analysis of the ecological crisis, not by advocating the abolition of nature or the dissolution of ecological problems (Hayward 1994), but rather by using the tools of critical thinking for deconstructing the ecological modernization theory (Blühdorn 2007a, b). The ecological modernization theory is regarded as a mere peace-keeping strategy because it proposes the old faith in infinite growth, or rather the resilience of democratic consumer capitalism, coupled with declaratory commitments toward sustainability goals (Christoff 1996; Blühdorn and Welsh 2008); however, the full acceptance of strictly environmental values is incompatible with the practices of modern capitalist consumer democracies, and this determines a no-way-out paradox for mainstream environmental politics (Blühdorn 2011). Hence, it becomes clear that there are no alternatives to post-ecologism, and that society needs to better explore the ways in which to make this transition at least bearable for the long term.

Blühdorn determines that environmentalism did not provide, as philosopher Andrew Dobson claimed (Dobson 2003a), a correct analytical description of the society, the utopian ideals able to lead human action were too vague, and the program for political action was never really convincing from a sociological perspective (Blühdorn 2006). Blühdorn (2000, 156) adds "[e] cologists believed that categorically valid ecological imperatives (ecological *morality*) or a set of rationally justifiable ecological values (ecological *rationality*) would provide them with a normative basis for a radical critique and reorganization of contemporary society. Yet in practical terms, both the eco-morality and the eco-rationality failed." Consequently, the belief that ecologism can rightly claim itself a political ideology and its status as master framework for economic, legal, welfare, security questions, and so on, to be reorganized in accordance with ecological principles, is clearly misleading because ecological issues themselves were actually reframed "in accordance with the established principles of the various societal contexts" (Blühdorn 2000, 23). Moreover, Blühdorn contests that post-environmentalists' interpretations of environmental thinking assumes, in a very modernist light, that irrespective of the fact ecologists talk a lot about pluralization, or different positions such as conservationism, ecologism, environmentalism, ecological modernization, and so forth, all tend toward unity and common goals. Quite to the contrary, the only way real post-environmentalism can introduce a truly innovative perspective is by supporting the democratization and the pluralization of the monolithic concept of nature (Blühdorn 2000), including its very abolition. It is, in fact, through the abolition of the idea of a single nature that it is possible to rethink it as a historical and cultural object, so that while it "remained impossible to determine the exact relationship between physical environmental change and the social perception of problems, [...] there is no direct connection between environmental physical conditions and public environmental anxiety" (Blühdorn 2000, 41). Blühdorn presents a strong constructivist interpretation, which is defended from realist attacks by counter arguing that the debate about the existence of an external world outside of the discursive domain is something for epistemologists but misses the point from a sociological point of view. Blühdorn affirms that constructionists accept the existence of physical reality outside of their control and "certainly their approach can only complement and never replace that of the sciences" (Blühdorn 2000, 48). However, despite the fact that an objective source of environmental concern is necessary, "there is no reason to believe that this basis necessarily needs to be located in the material world" (Blühdorn 2000, 50). Blühdorn's preference for constructivism is motivated by an in-depth exploration of the reasons why the environmental movement has achieved such modest results to present date. Ecologism is analyzed under the context of Frankfurt critical theory from Theodor Adorno's theory of alienation and modernity (1973) up to Beck's reflexive theory (1996). This analysis is complemented by Niklas Luhmann's social systems theory (1995) and leads to the conclusion that "contemporary society is structurally unable to solve the so-called ecological problem, [also] society *does not really have to solve it*, anyway, because this problem is no more than a specifically ecologist construction" (Blühdorn 2000, 152). Post-ecologism therefore demonstrates that both ecological morality and rationality are unnecessarily generated by ecologists, and the elaboration of environmental concerns are actually problematic. Consequently, Blühdorn's (2000) deconstruction of the ecological credo results in the following:

> [An e]cology [i.e. an environmentalism] without identity both in the sense that the formerly central issues of identify no longer play a major role and in the sense that post-ecologism is no longer identifiable as an independent discourse and policy. Post-ecologism is post-natural, post-subjective, post-ethical and post-problematic. It has abandoned the ideals of unity and inclusiveness and replaced them by plurality and differentiation. It is strongly individualistic and self–centred. It places the emphasis on on-going construction and innovation rather than conservation and preservation. It celebrates the present and reveals in momentary pleasure and excitement (172).[5]

Still, not all hope is lost and there exists the possibility for this theory without identity to rediscover itself, ideally where the principal ideology would include the abolition of nature, the abdication of the modernist subject and the end of eco-ethics. The egalitarian ideal of ecologism, and particularly its consideration of social inclusion among environmental problems, is very controversial. Surely everyone can agree upon the evidence that, if we are defining equality as the promotion of high standards of life for everybody, it results in ecologically disrupting outcomes. Consequently, Blühdorn affirms that post-ecologism needs to recognize the naturalization of exclusion (Sachs 1997), or to recognize that "in contrast to the ecological ideal of inclusiveness, the objective of contemporary policies of social inclusion is really to perpetuate the principle of social exclusion" (Blühdorn 2000, 167). In this argument, Blühdorn characterizes post-ecologism as a nonideological ideology which allows society to continue with business as usual by figuring something is done about the environment and that "ecological modernization thus promotes and facilitates the continuation of the established socio-economic practice, while at the same time confirms the belief that society is performing the ecological U-shape turn" (Blühdorn 2000, 198). Nonetheless, it is not ecological modernization that needs to be contested, but rather the social faith in its capability to overcome and overturn current values and behaviors. The last statement is, in fact, the essential paradox of environmental politics, this inherent will to sustain the unsustainable, or to keep infinite growth and wealth accumulation principles unchanged while marrying them with environmental measures and principles governing production and consumption practices. We are witnessing the resilience of democratic consumer capitalism advanced by post-ecologist politics of unsustainability. Obviously, post-ecologist politics are supported by a sort of tacit alliance between a broad range of actors all coalescent of the governance of sustainability, apparently regardless of the differences between producers and recipients of symbolic power (Blühdorn 2007b). According to Blühdorn, the focus on governance rather than on government allows us to include a plethora of actors all apparently engaged in finding solutions but none formulating the problems in a way that can challenge the constitutive principle of established order (Blühdorn 2011).

This generates an impasse determined by the concomitant acceptance of strictly environmental values, lifestyle, and social practices as a sort of declaratory hyper-ecologism and results in a profound inability and unwillingness to implement the required changes (Blühdorn 2011). From this perspective, post-ecologism is not a solution but rather a sign of the inability of modern society to perform a turn toward sustainability by merely reinforcing the current managerial and technological systems for securing the existing structures (Blühdorn 2011). The fundamental question of environmental thinking in post-ecologist era is not anymore "how can we change societal practices in a way that they become more sustainable," but rather "how [may we] sustain social structures and lifestyle that are unsustainable, i.e. how can we manage to sustain the unsustainable?" (Blühdorn and Welsh 2008).

Box 3.2   The post-ecologist construction of nature: the international campaign "El *Yasuní Depende de Ti* "

Ecuador's internationally recognized environmental campaign "El Yasuní Depende de Ti" clearly represents a case of social construction of nature in the post-ecologist context, which fully integrates in a single proposal conservationism with the critical stances of political ecology, the UN mainstream framework for ecosystems assessment, and the payment for ecosystem services processes.

The Yasuní National Park (YNP) is a richly bio- and ethno-diverse UNESCO World biosphere reserve in the Ecuadorian Amazonia. The park's capacities have been increasingly jeopardized as a result of the expansion of oil extraction activities in the most remote forest areas (Certomà and Greyl 2012). One of the extractive areas, the Ishpingo-Tambococha-Tiputini (ITT) block, corresponds to the area covered by the Yasuní ITT project (CDCA, 2011). In March 2007, the Ecuadorian government launched the Yasuní ITT proposal to suspend the extraction of crude oil from the ITT block. This was an estimated total of 900 million barrels, corresponding to 10 days of average global oil consumption and 407 million tons of saved carbon emissions (CDCA, 2011).[6] On the basis of the value of Certified Emission Reductions (CERs) in the European market during May 2009 (17.66 dollars per metric ton), the economic value of the saved emissions was estimated by the Ecuadorian government to be valued around 7.2 billion dollars (Government of Ecuador 2010). Without accounting in monetary terms for the protection of the bio- and ethno-diversity concentrated in the YNP, this is approximately the same amount the Ecuadorian state would have profited for the exploitation of the ITT block. The Yasuní ITT project team envisaged that the United Nations Development

Programme (UNDP) would gather pledges from international and national organizations, and international cooperation programs, among other individuals, and that international government economic support would also be received in exchange for bonds named Yasuní Guarantee Certificates, or CGYs.[7] Ideally, the money collected through the UNDP, along with a capitalization of a 7% benefit, would be reinvested into renewable energy projects, deforestation prevention and reforestation projects, ecosystems and biodiversity conservation programs, social development initiatives in degraded areas, and technological innovation (Government of Ecuador 2010; Larrea and Warnas 2009). The project was well received by both social and environmental movements, who applauded the project on its relation between non-extractive policies, and climate change and social justice initiatives.

In the end, however, financial targets were not met and the project closed, leaving NGOs looking to adopt alternative lobbying strategies for Yasuní preservation. Upon further investigation, however, the Yasuní ITT proposal itself was not without culpability. Originally presented as a new mechanism to prevent greenhouse gas emissions, considerable doubt existed on its real innovative character, as it might reasonably be argued this project was little more than another ploy to introduce some new product into the carbon market, such as the CGYs, rather than advancing alternative commitments to post-Kyoto regime, particularly the Clean Development Mechanism.[8] Despite the Ecuadorian government's claims that the Yasuní ITT initiative was not a sale of environmental services, but rather compensation for the loss in profits from oil exploitation, the effective realization of the project did not directly question the pillars of the capitalist model of energy production and consumption, and it did not explicitly address the geometries of power in the global economic system. Rather, this is a clear example of the substitution of dependence on foreign investments in oil exploitation for dependence upon foreign subsides in forest preservation. The isolation of wild nature areas, which are considered valuable and deserving of global conservation efforts, such as the ITT block, often leaves surrounding development to business-as-usual and does not produce a change in the collective mentality.

The constructivist interpretation of post-environmentalism, exemplary in its interpretation by Blühdorn's post-ecologism theory, offers some interesting suggestions for appreciating the current evolution of environmental thinking. However, as also discussed throughout this chapter, it presents a number of obstacles that weaken its argumentative strength.

The epistemological foundation of the discursive construction of nature, despite being disregarded by Blühdorn as not pertaining to the sphere of environmental politics, is of fundamental importance for appreciating the effects of the crisis of scientific *representation* over political *representativity* of environmentalism. Blühdorn's disinterest for the relationship between science and society is at the root of the ambiguous interpretation of the role of scientific data, whose supremacy in public debate, while at times contested, is acknowledged forthright, as scientific evidence has long been recognized to motivate public agency. The discrepancy becomes evident when this acknowledgment is flanked by the affirmation, in line with Luhmann's position, that environmental problems do not have a cogence in themselves because they are imaginary constructions of environmentalists rather than a piece of reality, and consequently have no real need to be addressed. It seems, however, hard to understand how, for instance, the consequences of the 1984 explosion of the Bhopal Union Carbide pesticide plant in India, causing the death and severe injury to more than 20,000 people, can be regarded as a mere invention of ecologists. Discourses, imagery, symbols, and narratives are powerful means of influencing public opinion and agency, but they can in no way be regarded, as even Blühdorn himself partly recognizes, as the only object of environmental claims. Moreover, Blühdorn's critique of the increasing relevance of governance, not government, networks in post-ecologism does not recognize that much sociological and anthropological research has already definitively argued that governance can also empower citizens and grassroots movements and activism in society.

In conclusion, while both the realist and the constructivist perspectives on post-environmentalism propose some relevant insights on the future evolution of environmental thinking, they still seem to lack an innovative understanding of human and environment relationship able to affect true innovation for thought and action, as displayed in Table 3.1. Chapter 4 will further explore this point.

Table 3.1 A synthetic comparison between Young, Shellenberger and Nordhaus, and Blühdorn's reference theories

Theoretical references	Current interpretations of post-environmentalism		
	Young	Shellenberger and Nordhaus	Blühdorn
Political Ecology	• Anarchist inspiration • Issue-oriented approach • Focus on social aspects of environmental issues • Technology friendly • Contestation of apolitical ecology	• Focus on social aspects of environmental issues • Technology friendly • Contestation of apolitical ecology	• Contestation of apolitical ecology
Realism		• Ecological modernization • Attention for sociopolitical data • Role of green business and communication • Progressive inspiration	
Constructivism			• Reflexive modernization • Sociocultural perceptions as objects of investigation • Critique of late modern capitalism • Universalism

Source: Author

NOTES

1. See for instance the *Inshore Fisheries Aggregating Devices*, WWF Australia project in the Solomon Islands (WWF 1995). This process is called essentialisation.

2. However, different from political ecology scholars, Schellenberger and Nordhaus' analysis is deeply connected to the US social and political context. Their aim is to revitalize progressive and liberal political forces by including environmentalism amongst them (while liberalism in other countries, especially across Europe, is at best understood as social democratic force rather than progressive). There is in fact a certain convergence among political scientists on the idea that environmentalism cannot be anything other than liberal (Arial Maldonado 2012).

3. Following other Critical Thinkers, Blühdorn uses the world "ecologism." Despite the fact that he does not devote particular attention to the difference between ecologism and environmentalism, the difference is actually defined in other relevant literature (Castells 1998; Hayward 2003). Political philosopher Andrew Dobson states that environmentalism proposes "a managerial approach to environmental problems, [and affirms] that they can be solved without fundamental changes in present values or patterns of production and consumption [while ecologism holds that] a sustainable and fulfilling existence presupposes radical changes in our relationship with the non-human natural world, and in our mode of social and political life" (Dobson 2003a, 365). For the purposes of this book, such differences are not very relevant, and the definition of environmental thinking is adopted to include the variegated world of those concerned with environmental issues.

4. As already mentioned, the terms ecologists and ecologism, rather than environmentalists and environmentalism (as preferred in this book), are adopted here to reference Blühdorn. The differences, however, are not relevant in this context, and the terms can be used synonymously.

5. Blühdorn's indistinct use of the word ecology, instead of environment, leads to the paradoxical affirmation that post-ecologism was an ecology without identity. Post-ecologism (or post-environmentalism) is clearly not an ecology because ecology is a natural science with its own status, heuristic processes, paradigms, and objects of research.

6. For Ecuador, this was comparable to 19 years of carbon emissions and the production of 107.00 barrels of oil per day for at least 13 years of full exploitation (Government of Ecuador 2011).

7. CGYs are also sold to private investors that develop projects in line with the Clean Development Mechanism guidelines, as established by the Kyoto Protocol under conditions not to exceed the total quota of annual emission

permits. However, because market-based revenues from the sale of certificates of avoided emissions are not currently recognized in the carbon market, a dedicated agreement is required.

8. The post-Kyoto regime derives from the recognition of the marginal results actually achieved in the implementation of Kyoto protocol, and the considerable criticism that followed. This new regime proposes stricter measures, particularly the UN Reducing Emission from Deforestation and Degradation (UN-REDD) Programme. Nevertheless, it too has been severely criticized, especially by indigenous groups for not clearly taking into account human resources and social issues (Larrea and Warnas 2009).

CHAPTER 4

Postenvironmentalism *beyond* Post-environmentalism

Abstract This chapter starts with the consideration that different post-environmentalist theories seem to be unable to provide an inspiring message for people engagement in environmental issues and introduces an alternative perspective based on the post-modern material-semiotic theory. This emerged from the seminal contribute of science sociologists and critical geographers which explored the constitutively heterogeneous characters of socio-environmental agents as both natural and cultural at once. From such a perspective, the chapter investigates how material semiotics can contribute to overcome existing interpretations of post-environmentalism, by challenging common understanding of the world ontology as well as mainstream epistemological perspective. The result suggests the need for a new gaze on existing forms of environmental commitment, which is here named as post-environmentalism (without hyphen) through which the whole, multi-layered, complex process of making and unmaking the world performed by hybrid assemblages is regarded as a political activity.

Keywords Postenvironmentalism · Latour · hybrid actors · heterogeneous networks

Both realist and constructivist post-environmentalist scholars have up to now deconstructed mainstream environmentalism by providing in depth analyses of its failings and weaknesses, which are what resulted in the loss

© The Author(s) 2016
C. Certomà, *Postenvironmentalism*,
DOI 10.1057/978-1-137-50790-7_4

69

of its once inspiring and powerful social and political capabilities. To many environmental commenters, Shellenberger and Nordhaus' theory of post-environmentalism represents an interesting example of the ecological modernization theory (Rootes 2008; Schlosberg and Rinfret 2008; Luke 2009), or rather of how post-environmentalism can be interpreted from a realist perspective. With help of natural sciences, Shellenberger and Nordhaus build their argument upon various socio-environmental data, or data relating to achievement in terms of environmental protection and restoration measures, funds appropriated in lobbying and campaigning, and other data concerning public opinions. Taking a post-structuralist approach, they contend that environmental issues are a matter of combined social understanding and political decision making and express urgent concern regarding the economic and political impediments that need to be overcome in order to define a more impacting marketing and communication strategy. This has generated considerable debate among both environmentalists and international scholars (Bate 1995; Meyer 2005; Blühdorn and Welsh 2007; Brick and Cawley 2008; Chaloupka 2008). For instance, a punctual and detailed critique by *Sierra Club* Executive Director Carl Pope described Shellenberger and Nordhaus' contribution as "unfair, unclear and divisive" (Pope 2005), and thus rather useless with regard to assisting in the development of any new comprehensive and effective environmental strategies. Pope argues that Shellenberger and Nordhaus' conclusions are not at all based in the information collected from the interviews they conducted, and that announcing the death of environmentalism does not seem to be a particularly helpful way in which to push the environmental struggle forward. Also contrary to Shellenberger and Nordhaus, Pope points out that environmentalist groups have campaigned and collaborated with industry and workers' unions for decades now, and he expresses considerable disapproval with the claim that modern environmentalism is no longer capable of dealing with the world's most serious ecological crises, and that is more likely to create defensiveness and resistance rather than foster progress. Pope contends that Shellenberger and Nordhaus do not offer any real alternative, if only that a moral commitment and change of values are required.

Contrastingly, Shellenberger and Nordhaus' critique, while not explicitly recalled, is strongly resonant in the voices of some key exponents of the environmental movement, such as *Greenpeace* founder Patrick Moore, who has stated that environmentalism movement has become a religion

(Spahl 2014). Moore supports a consensual, rather than a confrontational anti-establishment approach, which has been staunchly demonstrated by *Greenpeace*, and resulted in environmentalism shifting from a position focusing on both environment and the people, to environment only. Moore maintains that the biggest environmental problem is poverty because "[p]oor people cannot afford to clean wastewater, to clean the air, to plant new trees after cutting them down for fuel, etc. Poverty is a problem for the people and for the environment" (Spahl 2014). What environmentalism needs is to strengthen democracy, and to support the inclusiveness of all minorities and underrepresented people. Such a point also helps to advance the post-environmentalist theory, understood in this context as an attempt at debunking the elitist view and exclusivist approach of many environmental organizations (Alcantara 2013).

The academic debate pointed out the contradictory strategy proposed by *The death of post-environmentalism*, emphasizing market-based initiatives and broad public commitment (Kysar 2008) and defined the relationship between economic growth and the prioritization of post-material values as problematic (Beevers and Petersen 2009; Davidson 2009). Particularly the political promise of creating full-time jobs with high wages through economic growth fueled by massive investment into clean energy development is regarded as a kind of "Fordist–Keynesian compromise between capital, labour, and the state" (Buck 2012, 4) rather than a first step toward a post-Fordist model.

In 2011, Shellenberger and Nordhaus published *Love Your Monsters: Post-environmentalism and the Anthropocene*, a collection of essays detailing their own understanding of post-environmentalism, in which optimism for the possibilities offered by technological progress characterizes environmentalism to come. The authors aim to dismantle neo-Malthusian environmentalism by replacing it with new theories based on the power of human creativity and technological abundance. The "monsters" mentioned in the title are actually a metaphor for technologies, a topic more deeply investigated by sociologist and contributor Bruno Latour. By recalling his previous work, *Politics of Nature* (2004), Latour affirms that environmentalists have comparatively misunderstood Mary Shelley's *Frankenstein* as a warning against the dangers of technologies and should instead view it as a representation of how technology can become a monster only after it is rejected and abandoned. Equally, we are afraid of technologies because of the unexpected consequences they can generate, instead of working toward ameliorating them (Latour 2011).

Despite the hight quality of the contribute, overall, the message throughout *Love Your Monsters* reinforces Shellenberger and Nordhaus' theory that economic development and technological innovation means the salvation from the limits to growth (Ellis 2011; Sagoff 2011; Sarewitz 2011). The collection does not advance any brand new perspectives and only provides a support for the ecological modernization theory with some critical insights suggested by political ecology and the postmodern deconstructionist tradition. Most notably, reviewers point out that the collection "fails to recognize the context that enabled the technological progress of the past two centuries to occur—the rise strong property rights and market economies." (Bailey 2012, 3). And that this was thighly linked with a liberal perspective.

It should be noted, however, that *Love Your Monsters* also attracted the approval of several important environmental commentators, such as John Horgan. He effectively summarizes the core of the collection in terms of alternative choices:

> If we want more forests and more wild places, then we'll need more people living in cities and more intensive agriculture. If we want less global warming, then we'll need to replace fossil energy with clean energy, including a lot of nuclear energy. If we want to save places like the Amazon rainforest then we have to recognize that, over the next 50 years, a lot of the Amazon is going to be developed. The choices will come down to where we want development, and what we might save in the process (Horgan 2011, 2).

The very issue, Horgan comments, is not limiting the human footprint but deciding how to impact on the planet; the increase of greener technologies is now the only viable alternative to the deterioration of global environment and its increasing moral and financial costs. Shellenberger and Nordhaus, thus, according to Horgan, deserve praise for their optimism in economic development and technological innovation, which introduce a non-apocalyptic approach to environmental commitment (Horgan 2011).

4.1 From Post-environmentalism to Postenvironmentalism

On the emergence of post-environmentalism, Bruno Latour contends that while the critical analysis by Shellenberger and Nordhaus is timely and insightful, solutions are not very original or insightful, particularly the proposal for reinvigorating political emotion for development under green auspices. Latour (2008) does admirably note:

> [Post-environmentalism tries] to overcome the tragic consequences of bringing Nature into politics: in the name of indisputable facts portraying a bleak future for the human race, Green politics has succeeded in depoliticizing political passions to the point of leaving citizens nothing but gloomy asceticism, a terror for trespassing over Nature and a diffidence toward industry, innovation, technology, and science (2).

This statement implies that Shellenberger and Nordhaus questioned the epistemology of politics based on the idea of limits of human intervention, which impedes a real interaction with the nature. Latour (2008), in line with the postmodern interpretation of environmental commitment, suggests that environmentalism needs to recognize the multiple attachments, heterogeneous imbroglios, and mixing of human and nonhumans. This would push us toward a new stream of environmental thinking, capable of overcoming the boundaries of ontological categories.

So while post-environmentalist critiques of mainstream environmental thinking appear to be punctual and insightful, they rarely suggest any real groundbreaking views of alternative possibilities both in terms of the theory and practice of environmentalism to come. Shellenberger and Nordhaus' redevelopment under green auspices project, which is strongly resonant with the environmental modernization blueprint, requires emancipation from nature and further deepens the ontological distinction between society and nature (Latour 2008). Oppositely, Blühdorn's post-ecologist theory brings to light the raise and decline of mainstream environmental thinking and criticizes the realist approach as naively scientist. Such a constructivist analysis is fully internal to Critical Thinkers' debate, whose interest is almost completely attracted by the discursive communication practices, and notably fails to engage with the materiality of real life. Again, the idea that nature does not actually exist outside of social elaboration reinforces, instead of eliminating, the dichotomy between natural and social domain. Simply stated, both realism and constructivist approaches are unable to overcome the dualist interpretation of reality (Castree, MacMillan 2011) and perpetuate a fetishization of nature that deepens the gap between nature and society as an unbridgeable distance (Darier 1999; Descola 1996).

In such a seemingly hopeless landscape is there, or there can be, anything else, after the end of environmentalism so loudly proclaimed by post-environmentalism?

This "something else," I claim, can be called *postenvironmentalism*, meaning a transformation of environmental thinking which is not a mere opposition toward the past (as post-environmentalism was), but rather a new form of environmental thinking endowed with its own character. This will reconsider the long tradition of environmental thinking with fresh perspective and will builds upon its legacy. In so doing, postenvironmentalism can re-politicize environmental issues by understanding the complex and daily task of making and unmaking the world as a political activity which requires the mobilization of the social, environmental, and techno-scientific dimensions all at once, or the so-called natureculturetechnics category which is discussed in greater detail later in this chapter. While not blindly relying on technological solutions to address environmental problems, it devotes care and attention to technologies as coproducers of our world, such as in the context of Latour (2011). While it fully subscribes the pluralism of social understanding of nature that Blühdorn extensively describes (see Chapter 3.3), it does not fall into a solipsism that dissolves the reality of nature as a mere cultural artifact or defines environmental problems as the imaginary constructions of environmentalists. On the contrary, rooting in the political ecology tradition, it maintains the very existence of environmental facts generated by the interplay of science, power, society, and nature. Thus, this emerging postenvironmentalist theory fully participates in the constitution and functioning of global governance processes by striving to change the existing relations of power.

In order to fully appreciate the character and function of this new postenvironmentalist thinking, we need to literally and figuratively dig up the roots and subvert much of the previous thought that resulted in mainstream environmental thinking as unsustainable and ineffective in the first place (see Fig. 4.1). In line with the post-structuralist tradition, we need to deconstruct our epistemological understanding of the ontology of the world, the epistemological practices we use to understand it, and the functioning of the political practices aimed at changing it.

For example, the decrease in public commitment to environmental issues was in part generated by the progressive de-politicization of environmental issues advanced by mainstream environmental thinking in the hopes of making environmental protection an undisputed global priority (Hinchliffe 2007). This very observation lead Latour (2004) to affirm that

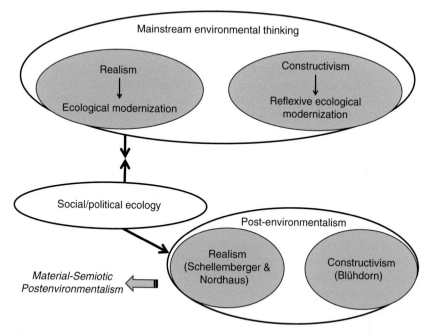

Fig. 4.1 Toward material-semiotic postenvironmentalism

Source: the author

"[d]espite what it often asserts, [...] nature is the chief obstacle that has always hampered the development of public discourse" (9). So in actual fact, in aiming to protect nature, environmental movements first and international environmental politics later, have unwittingly adopted a strategy of promoting a very unachievable vision of nature that ultimately results in the failure of their political struggle, because they rely on an apolitical concept of nature to support and further their causes. This has also generated the development of an impenetrable imbroglio of politics, nature, and knowledge that has endowed experts, notably natural and hard scientists, with the greatest power ever imagined, that is to make the mute natural world able to speak through *ad hoc* devices and procedures. Expert opinion is therefore profusely legitimated by the authority of nature itself and, at the eyes of mainstream environmentalism, this makes it possible to end interminable partisan debates by demonstrating that the truth emerges from interrogating reality itself. This entanglement between sources of knowledge and practices of power originates from the belief that the extraction of environmental issues from the factious and uncertain domain of sociopolitical disputes will have made it possible for experts to determine the truth about them, and thus to light the (right) way forward. However, it also generates the paradoxical situation in which the description of a nature in need of protection actually requires such a large intervention of experts and devices that nature itself almost does not exist anymore (Latour 2004).

The idea of postenvironmentalism as a new paradigm for environmental thinking, and not just merely another critique, has been in some ways previously suggested by Bruno Latour (2011), by some postmodern science, technology and sociology scholars, and some radical geography scholars (such as Stephen Hinchliffe, Sarah Whatmore, or Jane Bennet). My proposal develops greatly upon on their seminal insights and describes the emerging postenvironmentalism as material semiotic.

Material-semiotic approach allows us to overcome the distinction between realism and constructivism by bringing people, nature, technology, laws, ecosystems, institutions, and so on, together equally in the complex business of determining the fate of the world in which they commonly dwell. Through enacting heterogeneous actor networks in the constitution of the world, both addresses the problematic emancipatory stances of realism by uprooting and reversing the dualism problems, and the discursive closure of constructivism by attracting public attention on the practices materializing environmental thinking.

4.2   Hybrid Actors in Heterogeneous Networks

Much of traditional environmental thinking relies on the belief that ontological order is the original condition of world, and harmonious societies are those that, thanks to an accurate observance and reproduction of the natural laws, are able to reproduce it. As the classic German myth of Faust clearly seems to indicate,[1] the world was intrinsically ordered until modern hubris interfered with the primordial status. The subsequent modern obsession with the categorization, cleaning up, and purification of the world derives from the belief that different things and beings originally belonged to different domains; and allows control and domination over the nonhuman world. The progressive expansion of this power, however, also generated a deep sense of guilt when it was pushed over the nonhuman or natural world and made it a disenchanted and arid object of scientific investigation. Since the publication of Max Weber's *The Protestant Ethic and the Spirit of Capitalism* in 1905, many reputed Critical Thinker scholars and other social scientists, including Charles Taylor (1992), Anthony Giddens (1990), and Clifford Geertz (2000), have extensively expanded upon the description of the consequences of modern culture in terms of alienating individualism, the decline of the authentic quality of the collective experience, and the loss of the world's authenticity.

Material-semiotic scholars, however, suggest a different interpretation. Latour (2004) points out that order was not an ontological quality of the premodern world, and humans are not so sharply separated from the rest of the world. Distinction and separation are the results of the modernist attempt to categorize and control a world which is actually in chaos; consequently modern scientific and cultural practices did not mix beings and things that should not be mixed, exactly because categories, separations, classifications, and hierarchies have never been actual. They are generated from our epistemological, moral, and political frameworks of categorization and do not emanate from any preexistent ontological order. As a consequence, the mainstream environmental thinking relies on the assumption that environmental problems derive from the encroachment of different ontological domains that should be kept separate. On the contrary, Latour suggests that what we currently experience is not emancipation from nature (i.e., the separation of different domains), but the emancipation of nature (i.e., a more intimate connection between different domains) (Latour 2008). So while the modernist rhetoric describes a future characterized by less and less attachments, what is actually happening is a world where people are becoming more and more entangled in the blending of human and nonhumans actors (Latour 1996; Fall 2014).

> **Box 4.1 Dissolving categories: forest, savannah, and indigenous categories of nature**
> The fallacy of categorization and the misinterpretation generated by dualistic thinking on such issues as environment and culture, human, and nonhuman relationships, the natural and the artificial, and so on, emerges clearly from critical studies on indigenous peoples' relationships to nature. Latour (2005), for example, explains his critique of

the common idea that indigenous peoples have a more direct relationship with nature because of their respect of the pristine natural order and traditional values, and states that

> [w]e must ... disappoint those who imagine that other cultures will have a richer vision of nature than our own Western version. It's impossible to blame those who share such illusion. Countless words have been written ridiculing the miserable whites who are guilty of wanting to master, mistreat, dominate, possess, reject, violate and rape nature. No book of theoretical ecology fails to shame them by contrasting the wretched objectivity of western with the timeless wisdom of 'savages' who for their part are said to 'respect nature', 'live in harmony with her', and plumb her most intimate secret, fusing their souls with those of things, speaking with animals, marrying plants, engaging in discussions on an equal footing with the planet (42).

Indigenous cultures have been generally regarded as having closer relationships with their environment than even western environmentalists themselves. This view has been perpetuated by the western world's fascination with the exotic, as exhibited by the tremendous popularity of expeditions to the orient, India, and even the colonial territories since the seventeenth century, and the imagery of a primitive harmony in far-away places untouched by modernity. The myth of immediate empathy between indigenous peoples and the natural world was then perpetuated by the environmentalist belief that this original ecological wisdom has been lost forever for western and particularly modern people (Milton 1996). Deep ecology, which claims to be the more radical form of environmentalism, is indeed only the more radical form of this imbroglio. In fact, environmental thinking does not only stress the separateness of the natural and human domains and claims to privilege the natural one, but it also proposes to learn from non-western cultures how to do it. This view, obviously, ignores that which is often referred to as harmony between indigenous peoples and nature and does not derive from a particular sympathetic relation but from a different categorization. Upon closer analysis, Latour notes that comparative anthropology reveals that non-western cultures never adopted nature as a category per se, but they simply have never been interested in nature in the same way as

4 POSTENVIRONMENTALISM *BEYOND* POST-ENVIRONMENTALISM   79

the western world did. It was western philosophy that transformed nature into an object of concern and constantly dragged it into the definition of the political and social order (Latour 2005). Non-western cultures do not necessarily have a higher consideration of nature category than that of the West because nature is not regarded as a standalone ontological domain, opposed to the culture one. As Latour (2005) states "[t]he difference no longer lay in the savages' not treating nature well, but rather in their not treating it at all" (44). This means that in many indigenous cultures there is only a single category including associations of humans and nonhumans.

Socio-agronomical research confirms Latour's hypothesis. Such as in the case of desertification in former French colonial West African territories, French colonizers originally perceived desertification as an effect of indigenous mismanagement of the land. Sometime later, however, post-colonialist studies provided a different explanation of the phenomenon and affirmed it to be associated with the colonial and neocolonial marginalization of smallholders and pastoralists (Adger et al. 2001). James Fairhead and Melissa Leach's 1996 publication, *Misreading the African Landscape: Society and Ecology in a Forest-Savanna Mosaic*, expands significantly on this point by explaining that, for instance, French administration of Guinée since the end of the 1990s century, assumed forest patches to be relics of an original rainforest which once fully covered the landscape. French administration thought that the fire-setting practices adopted by local inhabitants converted the forest into savannah and threatened the possibility for agriculture. Subsequently, repressive policies aimed at correcting this supposedly destructive land management practice were adopted. At the time, Kissindougou's villagers in the South of Guinea suggested, however, that forest islands are not relics of a destructive practice, but a legacy left by ancestors of the inhabitants of the savannah. This opened up a completely new consideration of the Guinea landscape. This case exemplifies how the use of nature/culture dichotomy in anthropological studies is a reflection of the western metaphysic but is inadequate in interpreting non-western worldviews. Indeed, in many non-western societies such a distinction is nonsense. Kissindougou's rural inhabitants do not associate the presence of forest with a natural state because the presence of trees is the consequence of human agronomics; while savannah is actually associated with the state of nature. By adopting the nature/culture dichotomy as a means for interpreting ecological phenomena, forest degradation has been understood in terms of social dysfunction and as the effect of a breakdown of the socio-ecological equilibrium (Fairhead and Leach 1996). In western views, the use of land and vegetation in the absence of specific regulation and technologies is seen as potentially degrading, but obviously this contradicts the experience of the Kissindougou people, who believe that undeveloped land tends to be savannah, while cultivated lands become forest. Comparative anthropology, thus, supports the claim that indigenous people are not intrinsically closer to nature, and therefore that nor are western people distant from it. The cause of all the misinterpretation is the false distinction between the natural and cultural domain, which is what has lead western people to complain about their distance from nature, and to believe other culture to be closer to it. Latour (2005) notes:

> Westerners believe that they are detached from nature because they have forgotten the lesson of other cultures and live in a world of pure, efficient, profitable, and objective things; and ... other cultures believe that they had lived too long in the fusion between the natural order and the social order, and that they need finally, in order to accede to modernity, to take into account the nature of things 'as they are' (46).

Positivist culture, including mainstream environmental thinking, has maintained the distinction between ontological domains as unbridgeable in order to avoid any mixing which might result in the creation of hybrids of nature and culture, as the repulsion for hybrids represents one of the most distinctive traits of modernity that differs from pre-modernity exactly by the maniacal attempt at performing ontological purification, (Latour 1993b). Paradoxically, modern scientific and laboratory practices are flooding the world with hybrids of any sort, including vegetable species equipped with genetic features to survive in extreme conditions, human prostheses, artificial intelligence etc. All of them part of a single "natureculturetechnics" category (Law 2004). Geographer Steven Hinchliffe notes that hybrids are unstable formations emerging from the crisscrossing of in-becoming entities (Whatmore 2002), and they bring to the forefront evidence that differences among entities are not differences *in kind*, but rather differences *in degree*

(Hinchliffe 2007). Everything and everybody can be regarded as a hybrid generated by the entanglements of nature, technology, and society because hybridization does not imply everything must be part of everything else and does not level all the differences in an undifferentiated mixture. There are, in fact, possible and impossible mixes and matches, which are determined by structure, reactivity, and the openness of the involved things and beings, as well as partial forms of hybridization (Hinchliffe 2007; Fall 2005). This perspective stems from Deleuze and Guattari's (2002) description of the world as composed by a myriad of ceaseless attempts at establishing connections through an exchange of physical matter:

> [Y]ou become-animal only if, by whatever means or elements you emit corpuscles that enter the relation of movement and rest of the animal particles, or what amounts to the same things, that enter the zone of proximity of the animal molecule (275).

Following a similar argument, by dismissing the Weberian description of modern world as a place of disenchantment (Weber 2005) political scholar Jane Bennet describes the networking as a source of enchantment per se (2001). While the modern world has been generally represented as a place of alienation, disenchantment, abstract reason, and bureaucratic control, she describes the contemporary world as a place where the marvelous emerges everyday exactly from the practices of hybridization. Bennet (2001, 156) refers to enchantment as the condition of surprise for unexpected encounters, and "a feeling of being connected in an affirmative way to existence; it is to be under the momentary impression that the natural and cultural worlds *offer gifts* and, in so doing, remind us that it is good to be alive."

Together with the beauty of nature, also hi-tech products or the magic of money, the dream of nation, the ritual of finance, and similar can be a source of enchantment, because enchantment exactly resides in the discovery of material complexity (Dube 2002). Such an enchanted view of materialism originates in the molecular flows blending nature and culture in networks of transient beings that interweave heterogeneous elements, for example humans, nonhumans, and machines (Bennet 2001).

This perspective on hybridity helps, thus, in overcoming the dualistic impasse of the realist approach, and of the discourse orientated constructivist analysis. In addition, the account of reality as an intricate and dynamic entanglement of the physical and symbolic matter of the world helps to introduce a material-semiotic perspective (Arias-Maldonado 2015), which may shed new light on the meaning, the practices, and future of environmental thinking. The ontological considerations proposed by the evocative and intentionally provocative language of the material-semiotic approach discloses the intrinsic interrelatedness of domains traditionally regarded as distinct and offers new epistemological perspectives. This approach also characterizes a large number of postmodern attempts at overcoming the dialectic between realism and constructivism by focusing on the structuration and functioning of heterogeneous networks of humans, nonhumans (such as animals, plants, and ecosystems), and more-than humans (such as machines, rules, technologies, and procedures) (Hinchliffe and Whatmore 2006). While in some spiritualist or radical ecocentric philosophies, the entanglements of humans and nonhumans is already envisaged, the material semiotic introduces more-than humans too, principally because machines (e.g. computers), devices (e.g. sensors), and procedures (e.g. software) play such a key role in the proliferation of hybrids, resulting in a highly intimate connection of human with nonhumans (Haraway 1991). While the material semiotic requires an appreciation of the material weight of discourse, symbols, and ideas, it also importantly populates the public space with nonhumans or more-than-human things and beings that are not passively affected by human action but are endowed with agency capability. Nonhuman actors are considered as able to handle signs, so that the world "semiotic" should be taken, in a broad sense, as expressing the symbolic dimension of the world generated via material practices (Mol and Law 2002). This means that they do not merely resist and react to human stimuli but are rather provided with the inherent liveliness of autonomous existence (Bennet 2010). Thus, people, things, theories, forces, and various kinds of entities interact in the form of networks and provoke the material form of the world. In fact, the basic assumption of the material-semiotic perspective is that everything or every being is materially and discursively generated from and located in a network of relations (Law 2008).

Material-semiotic research started with the recognition of the scarce consideration of the roles of these connections (Leigh and Star 1991) in the sociology subfield of science, technology, and society studies (Law 1991). Law (1991) states that society is put together by a variety of heterogeneous means, and that "whenever we scrape the social surface we will find it is composed of network of heterogeneous materials" (10). In order to appreciate the reciprocal constitution of society and technologies, the material-semiotic prescribes to "follow the thing" in the process of both self-production and society production. In this context, from an epistemological perspective, knowledge itself is described as a relational effect because the objects of knowledge do not passively wait to be mirrored in accurate representations (Olesen, Markussen 2007). Rather, it is a practice able to interfere with other practices to generate different forms of materiality (Mol 2002). Its content can obviously vary as a function of social contexts; however the recognition that standards for good knowledge may differ between societies (also referred to as epistemological relativism), does not mean that standards do not exist. It does not advocate for relativism, but rather it helps us to beware of absolutism. In this context-dependent process of knowledge generation, politics obviously plays a key role. This point has been extensively addressed by actor-network theory (ANT).

In the panorama of material-semiotic-inspired theories of the early 1980s, Latour, in collaboration with sociologists Michel Callon and John Law, proposed a theoretical approach for disentangling the interplay of modern knowledge and politics generation processes, called ANT (Latour 2004; Law 2008). Law describes ANT as a broad corpus of different contributions all having in common a network-orientated understanding of the social phenomena (Law 2008). It is a case-based theory investigating how different things and beings do and do not assemble, how they turn into social actors, the consequences of their heterogeneous and hybrid formation, and how their assembling affects the spatial and material constitution of (power) relations working upon different geographical scales.[2] These relationships, particularly power practices, can be grasped in their full implications when we consider the ways in which nonhuman and more-than-human actors have made society durable and reproducible (Latour 1991) through particular material arrangements (Law and Hetherington 2003). For instance, Latour (1993a) points out that microbes in a number of circumstances play a crucial role in the composition of the social:

> Society is not made up just of men, for everywhere microbes intervene and act....In all these relations, these one-on-one confrontations ...[O]ther agents are present, acting, exchanging their contracts, imposing aims, and redefining the social bond in a different way. Cholera is no respecter of Mecca, enters the intestine of the Hadji; the gas bacillus has nothing against the woman in childhood, but it requires that she dies. In the midst of so-called 'social' relations, they both form alliances that complicate those relations in a terrible way. (233)

ANT provides the terrain for a material-semiotic, sociopolitical perspective to emerge not as a human prerogative, but as a distributed possibility

(Braun and Whatmore 2010). It is interested in how power is exerted by some chain of agents in order to produce specific configurations that make other actors behaving. As Stanforth (2006) notes:

> Those who are powerful are not those who hold power in principle but those who practically define or redefine what holds everyone together. This shift from principle to practice allows the vague notion of power to be treated not as a cause of people's behavior but as a consequence of an intense activity of enrolling, convincing and enlisting. (39)

While ANT has been criticized because it seems to flatten the agency on networking, it is nonetheless clear that it is not interested in providing a comprehensive theory of the political, but rather aims to show how action is not a straightforward effect of agents' intentionality because of the need to be equipped with specific relationships to perform in the public space. A number of networks assemble around a matter of concern, or a disputed state of affairs (Latour and Weibel 2005). They do not form an assembly in the traditional sense, such as a dedicated situation in which they directly face one another's idea, but they are politically active in practically dealing with peculiar forms and functioning of the world, so that they are said to *become* an assembly. Assemblies provided with political subjectivity in this context are referred to as *collectives* and their emergence signifies the passage from the ontological dimension of assemblages to the political one where they are able to take action in the public space. Material-semiotic theorists have diversely approached the development of a standard definition for the political mechanisms underpinning material politics. As Law (2008) describes:

> Haraway uses tropes – most famously the cyborg – that interfere with the undermined politically and ethically obnoxious realities. Latour talks of ontopolitics . . . and of 'parliament of things' where what is real, and how it might live together, are provisionally determined .... [And] Mol talks of ontological politics .... There are no general solutions (156).

By opposing a strong object-avoiding tendency in political philosophy (Bingham and Hinchliffe 2008), material semiotic brings back to the forefront of public debate the thing matter, which was thrown out from political arena as trivial and passive. The resulting "thing-orientated politics" (or *Dingpolitik*) (Latour and Weibel 2005) can be regarded as a reaction toward the inability of traditional democratic processes to deal with many crucial contemporary issues that require the transgression of ontological boundaries, and a new way in which to consider how a broad variety of actors contribute to shaping the world in which they all commonly dwell (Marres and Rogers 2005). A thing-orientated political process, explains sociologist Noortje Marres, stimulates public engagement by assembling a collective around problems that are complex, such as the creation of a park, the establishment of a factory building, or the ratification of a law against carbon emissions, and it requires the public in whole to manage it because individual actors lack the power to singularly influence issues (Marres 2005). Contrary to the classic assumption that political assemblies are created for addressing all present and future political issues that may arise, material politics affirms that that members of a collective may temporarily share the common condition of being affected by a particular state or affairs. Involved actors are not necessarily in agreement but are linked together by the mutual interest on the fate of a specific issue and not by adhesion to common ideologies, values, or principles. Antagonistic relations and disagreement are thus the norm in democratic contexts that are able to grant space to hostility by defusing its destructive potential.[3] Marres (2005) describes this in the following scenario:

> Farmers in Kansas and vegetarians in Europe, people with HIV in Sub-Saharan Africa and the employees of pharmaceutical companies in the North: they are involved in a dispute. They disagree about such fundamentals as whether GM food or AIDS drugs qualify as public affairs—i.e. issues that are to be subjected to scrutiny and concern by the broader public. They disagree as to which institution should adopt the affair, let alone, how.... Emergence of a public affair must be understood as an opportunity for disagreement. (8)

This opportunity requires the use of the most important—despite often forgotten- listening capability, which subverts the traditional primacy of speaking capability. As Dobson (2010) notes, from Aristotle onward, politics has usually been associated with the capacity to speak and communicate judgments, and since this ability is a peculiar feature of human beings, nonhumans are thus excluded from the political sphere. However, "if the political subject has to be a speaking being, how can the putative subjects of green politics *be* political subjects?" (Dobson 2010, 8). Environmental issues urges to grant those other than in the current population (meaning future generations, animals, ecosystems, and machines) the higher political consideration. However, a simple increase of spokespersons representing them in the parliamentary arena does not automatically transform politics itself as access to discussion will continue to be limited to humans alone. Contrary to this view, Latour suggests that all nonhumans and more-than humans can indeed enter the political arena and but he explains:

> I have not required human subjects to share the right of speech of which they are so justly proud with galaxies, neurons, cells, viruses, plants, and glaciers. [...] I have simply recalled what ought to be taken as self-evident form now on: between the speaking subject of the political tradition and the mute things of the epistemological tradition, there always was a third term, *indisputable speech,* a previously invisible form of political and scientific life that made it possible sometimes to transform mute things into 'speaking facts', and sometimes to make speaking subjects mute by requiring them to bow down before nondiscussable matters of fact. (Latour 2004, 68)

When granting speaking capability to somebody or something, we acknowledge it to be a witness. Scientists design processes that allow both human and nonhuman actors, or "actants,"[4] to act, in order to become witnesses in the mediation processes that transform scientific into sociopolitical representations. In fact, it is clear that nonhumans cannot advance their preferences in discursive form, but laboratory practices involving both humans and more-than-humans need to invent speech prostheses, or instruments and tools to give voice to mute things and, at the same time, make speaking beings listen (Latour 2004). Some may argue that this would again confine nonhuman actors to a subordinate level, but when considering the things that cannot speak for themselves (as there are for example no talking mountains or offshore platforms, and no speaking theorems), it can be argued that neither human nor nonhuman beings really speak on their own, as they all require the mediation of scientific, symbolic, and political procedures and devices to allow them to actually be able to communicate. Technological devices cannot thus be considered as communicative tools that translate in understandable discourses the voice of water river fluctuations, ozone depletion, waves in the air, and so forth. It goes without saying that these speaking devices are never completely adequate. A translation is after all just that, a translation. Something almost always invariably gets lost, modified, or added, or the focus might become distorted. Translation implies a sort of betrayal and does not only domesticate the words but the ontology itself. However, as philosopher Andrew Barry (2001) suggests, even the most accurate scientific investigation cannot provide us with a true representation, or translation, of the world that is completely faithful, as it requires a technical intervention that necessarily abstracts the investigated object from its context. Consequently, public issues, such as environmental issues, are necessarily elaborated on the basis of partially distorted representations. Yet despite its imperfections, the translation process not only makes science to exit laboratory, it also gets the outside world entering laboratory and shows how science and politics adopt similar mechanisms, procedures, and devices.

4.3   Toward a Material-Semiotic Postenvironmentalism

In the very first pages of this book, I suggest that environmental thinking is not dead, but that it is transforming into something new—something *radically* new. As defined in Sect. 4.1, "material-semiotic postenvironmentalism" signifies that, despite transcending, contesting, and innovating the tradition of environmental thinking, it nonetheless inherited its legacy and builds on the consequences of institutional environmentalism mainstreaming, as well as on the radical strength of political ecology and on the critical positions of post-environmentalism, as shown in Fig. 4.1. But it also goes beyond, and in doing so it subverts many conventional understandings related to what environmentalism is about. Most importantly, material-semiotic postenvironmentalism is not something *to come*, as it already exists because, regardless of the institutional, philosophical, or academic crisis of environmental thinking, people and their nonhuman counterparts continue to search for their own way to practice what they believe to be the right way for inhabiting this world. That is to say, postenvironmentalism is as much a practice as it is theory, performed in local places by social agents that are not necessarily interested and knowledgeable of environmental thinking disputes. This also helps to explain the frequent use in the media of inappropriate words and categories of traditional environmentalism when reporting innovative practices, which could be easily ascribed to the domain of postenvironmentalist (Santolini 2012).

The increasing diffusion of postenvironmentalist practices is easily understandable. For quite some time now, social agents worldwide have faced the everyday evidence of their progressive entanglements with diverse hybrid materialities in a world that is increasingly mediated, produced, enacted, and contested *through* technological, cultural, and ecological networks (White and Wilbert 2006). This complex but fascinating postmodern world is described better than any other by material-semiotic approach, and some of its key concepts are of particular help in disclosing the character and functioning of (material-semiotic) postenvironmentalism. Consequently, the findings of various material-semiotic scholars have been adopted in the analysis of some of the most disparate environmental issues, ranging from the very specific such as the papaya trade (Cook 2004), to more broader and innovative themes such as the problem of electronic waste (Gabrys 2011; Parikka 2011), and including more traditional ones dealing with the formation of environmental regimes such as the convention of biodiversity (Bled 2010). While in all of these cases scholars offered new perspectives on investigated issues, very rarely is a general description of environmental thinking transformation actually attempted. This is most likely due to ANT's predisposition of avoiding generalization and focusing on single issues on a per case basis. Nonetheless, I believe some general framework indications would allow a better appreciation of the potential impact of the material-semiotic perspective on environmental thinking and its postenvironmentalist outcomes. Chapter 5 will thus discuss some of these perspectives in greater detail and provide examples based on the following consideration of material-semiotic treating of environmental issues.

As previously mentioned, a material-semiotic perspective requires the placement of materiality and space at the core of public debate and suggests that, always and everywhere, speaking of politics means speaking of particular aspects of the relationship between environment and its dwellers. As Latour (2004) laconically states, "[t]here has never been any other politics than the politics *of* nature, and there has never been any other nature than the nature *of* politics" (28). This means that all forms of political thinking have an environmental dimension, which largely overwhelms the space conventionally devoted to environmental issues as mere sectorial interests. The focus on materiality helps to debunk the traditional questions of environmentalism and to devote necessary attention to previously ignored linkages. For instance, geographer Nick Bingham explores how the seemingly endless debate on genetically modified organisms (GMOs) can be approached differently by giving voice to the marginalized life forms involved, such as bees, butterflies, and bacteria, and to the biopolitical questions they raise. As Bingham (2006) points out, this approach helps to demonstrate, for example, that

[t]he unique transnational migration of the Monarch butterfly, the endless round trips of the honey bee, the slow, slow action of a soil microorganism,

the biotechnologically protected growth of a GM crop even, all these life-form ^ narrative ^ trajectory ^ things" are the "polymorphous spacings […] by which the world appears each time according to a decidedly local turn [of events]" (Bingham 2006, 316). (492)

New technologies, such as genetic technologies, do not simply fall from the sky into an empty world, and digging deeply into the complex relationships of our world makes evident the inadequacy of our political mechanisms limited to the exercise of a human-centered politics. It suggests the need for an issue-orientated democratic articulation of the public debate capable of providing things-issue the attention they deserve and serves to demonstrate that as "we have became able to acknowledge and articulate that as individuals we are always (already) at once surrounded by objects of various kinds, participants in communities of practice of various kinds, and immersed in activity of various kinds" (Bingham 2006, 496).

Making environmental issues public reveals to be a much more complex affair than merely introducing them in the classic political arena. It calls for the recognition of a relational ontology in which space and place play a major role as *constituted and reconstituted* through political activity (Featherstone 2008). In fact, the possibility for political articulation of environmental issues is provided by the spatialization of different ways of imagining and practicing heterogeneous coexistence. Human, nonhuman, and more-than-human actors all commonly produce forms of political life, animate the public sphere, and provide a mutable shape to the social domain with previously unthinkable configurations (Featherstone 2008). Public life thus is the effect of a common, everyday way of dealing with things. As Marres (2012) suggests, those concerned with environmental issues are looking for ways to account for the capacity of nonhuman entities such as trees or power plant to mobilize the public, not that much by extending the traditional participation processes to other than human, but rather by allowing a form of *material* participation, which is accomplished through the deployment of specific technologies, settings, and things.

In 2005, the Karlsruhe Center for Art and Media hosted an exhibition entitled *Making Things Public: Atmospheres of Democracy*, a large exposition on the creation of "the public" and its political expressions. On this occasion, a large number of scholars explored the meaning and functioning of an issue-orientated democracy by building discussion upon the assumption that collectives emerge when disparate actors gather around specific matters of concern. Through the description of a large number of real-life examples, they focused on the relevance of listening, the process of assembling of heterogeneous collectives, and the crucial role of mediation tools. For example, an ethological study on sheep from Yorkshire, UK, demonstrated for instance how actors engaged in a public dispute do not necessarily speak for themselves, but are rather *made to speak*, thanks to dedicated scientific tools that accurately report their opinions about the transformation of their own place and make them visible (Despret 2005). When the behavior of sheep is observed for a long time, their communication practices emerge quite clearly. Ad hoc tools and procedures employed specifically to detect these forms of communication will very clearly demonstrate that animals certainly do have preferences and opinions, and that they even go so far as to rank these options too (Despret 2005). The possibility to discover, or to invent, appropriate devices to listen nonhuman voices was, for instance, also explored in another article about the presence of wolves in the French Alps (Mauz and Gravelle 2005). While wolves had populated the mountains for decades, for a long time their presence had simply gone unnoticed. Nonetheless, their presence *practically* conditioned the surrounding environment in terms of the presence of other animals, smells, soil chemical reactions, power relations in the ecosystem hierarchy, human activity in the woods, and so forth. The wolves' presence in the French Alps also affects the behavior of a broad range of social actors, such as hunters, government administrators, tourists, watchdog groups, guide dogs, farmers tending to flocks, and so on. However, the simple existence of something is not enough to count it as a social actor, particularly if its existence is silenced, and no effort is made for listening to its voice. Marking the presence of wolves in the French Alps through dedicated tracking and reporting processes have made it possible to weigh their political relevance in terms of context modification (Mauz and Gravelle 2005).

A further evocative account of nonhumans taking part in the public debate is provided by Bennet in her 2010 work *Vibrant Matter: A Political Ecology of Things*. Bennet explores the ontological and political consequences of rejecting the idea of matter as passive and inert stuff, for a description of the vibrant materiality that not only resists or impedes human agency but also acts as quasi-agent itself. Building on Spinoza's theory that that "conative bodies strive to enhance their power of activity by forming alliances with other bodies" (Bennet 2010, x), Bennet devotes considerable attention to the analysis of the micropolitics through which "ethical sensibilities and social relations are formed and reformed themselves political" (Bennet 2010, xii) and explores the vibrancy of the disparate bodies, such as the Baltimore litter, Darwin's worms, and the not-quite-bodies of electricity, stem cells, and so on. Most notably, Bennet attempts to describe the political ecology of vibrant matter by claiming that it is by becoming a member of the public, or member of society, that one becomes a political agent. Starting from the consideration of agency that Darwin's worms are intelligent improvisations responding to external problems, Bennet suggests broadening the domain of those counting as agents by endorsing Dewey's description of a public as "a confederation of bodies . . . pulled together not so much by choice (a public is not exactly a voluntary association) as by a shared experience of harm that, over time, coalesces into a 'problem'" (Bennet 2010, 100). A common problem affecting social agents, despite not all agents being affected equally,[5] is exactly the matter of concern that turns network into assemblies. As Bennet (2010) explains:

[Such] problems give rise to publics, publics are groups of bodies with the capacity to affect and be affected . . . A public is a cluster of bodies harmed by the actions of others or even by actions born from their own action . . . [and] harmed bodies draw near each other and seek to engage in new acts that will restore their power, protect against future harm, or compensate for damage done- in *that* consist their political action, which, fortunately or unfortunately, will also become conjoint action with a chain of indirect, unpredictable consequences (101).

It is important to note that any given response to a problem is less the result of a deliberation than of "fermentation" of the various proposition and energies of the affected bodies.

Mediation tools, such as files, diagrams, archives, and records,[6] allow non-humans and more-than-humans to take part in the public debate by multiplying the context for political decision, or—in Latour's words (2004)—the number of parliaments. Where a public gathers around a matter of concern, for example laboratories, churches, ecosystems, tribunals, and so on, and use different forms of representation to advance their opinions, this is considered a parliament (Stenger 2005). Nonetheless, the broadening of the political arena does not mean that nonhumans can vote, or serve in the traditional parliament or performing any other activities conventionally considered as distinctively political. Rather, this more radically stresses that those activities are not the only ways to perform politics. For instance, it has been already considered in Chapter 1 how conventional political mechanisms, procedures, institutions, and theories often prove inadequate in dealing with environmental issues. The activity of these dispersed parliaments consist of continuous, multiple, and material negotiations through which assemblies address the specific issues they care about in discursive and nondiscursive form, by involving intentions, opinions, and motivations sometime far beyond the human capability required to detect them. Negotiations take the practical form of material engagement, including the enrolment of allies in the network and deploying powerful relationships to maintain them (Stanforth 2006). The public relevance of nonhuman opinions decenters social agency from the logocentric arrogance, which restricts humans alone to the status of sociopolitical actors (Whatmore 2000). For instance, Barry (2001) analyzes a case of maintaining the operation of a chemical plant whose polluting activity is far from being considered a mere technological issue. The situation involves the inhabitants of the local territory, and to a certain extent the inhabitants of the greater surrounding geographical area, capital investors, filter mechanisms, workers, chemical and toxic waste, devices and norms for pollution control, and global market knowledge. Consequently, it is difficult to determine the perimeter of the issue itself (Barry 2001), and around the issue of keeping the polluting plant in operation, a collective emerges and gathers together humans (including citizens' associations, experts, landowners, bureaucrats, inhabitants, and workers), more-than-humans (the plant itself, administrative bodies, legal rules, security devices, production chain, and the media), and nonhumans (soil, ecosystems, animals, surrounding water basins, chemical atmospheric compositions, and the wind).

In general, the whole, multilayered, complex process of making and unmaking the world is regarded as a political activity. This, obviously, may well involve words but is not discursive in kind. When heterogeneous and hybrid assemblages gather in the form of assemblies, they endorse a number of representational techniques able to put people and things in energetic, synergic, and sometimes conflictive relations. By adopting a conventional definition of politics, these techniques are not immediately understandable as political; however, by enlarging the definition of politics to include the interaction of different agents in the public forum, they can manifest their intrinsic political character and allow disparate events to be brought into the public debate as expressions of specific positions about certain issues, for example climate change, animal and nature conservancy, or protests over transportation infrastructure development. Thus there are things that can bring dispersed geographies together, techniques that embody political engagements, devices that incorporate ontological descriptions, and places where political decision are actively taken, still unnoticed.

So far, we explored how material-semiotic scholars have approached environmental issues by focusing on some relevant theoretical innovations provided by the investigations of the networks assembling around specific environmental issues. However, much work needs to be done in order to establish new material-semiotic interpretation of environmental thinking that can be widely understood, conceptualized, and formalized. Chapter 5 will thus provide some suggestions on achieving this goal and will consider if a material-semiotic perspective can help in detecting how the death of environmentalism is turning into the birth of a new and forward-looking postenvironmentalism.

NOTES

1. Remarkable literary transposition of the myth has been written by Christopher Marlowe (*The Tragical History of Doctor Faustus*, 1604) and Johann Wolfgang von Goethe (Faust, 1808-1832).
2. In a later contribution, Latour explains ANT is not a theory, or a frozen structure instantaneously accessible without deformations, but rather a provisional proposal that is intended to designate a dynamic structure constantly regenerated by transformations and translations, summing up local and practical interactions. ANT in fact does not explain *why* something happens, rather *how* events are brought into existence by a number of assembling relations (Latour 1999).
3. The role of antagonism in politics has been explored by Chantal Mouffe and Ernesto Laclau's 1985 article on radical democracy. By building upon this assumption, radical democracy theory opposes the reflexive modernization (Mouffe 1998) in which democracy becomes dialogic in the attempt to overwhelm, at the same time, the opposition between progress and tradition, and the opposition between left and right without clashing.
4. Material-semiotic scholars derived the definition of actants from Algirdas Greimas' (1986) definition of an integral structural element upon which the narrative of tales resolves. To Latour (2005), the term an "actant" is a source of action; both humans and nonhumans, that can do things, have sufficient coherence to produce effects, alter the course of events, and make a difference.
5. Bennet (2010) explains "Persons, worms, leaves, bacteria, metals, and hurricanes have different types and degrees of power, just as different persons have different types and degrees of power, different worms have different types and degrees of power, and so on, depending on the time, place,

composition, and density of the formation. But surely the scope of democratization can be broadened to acknowledge more nonhumans in more ways" (109).

6. Notably, mediation tools such as voting machines, electoral procedures, parliamentary rooms, the media, and so forth are crucial also in conventional understanding of political organization.

CHAPTER 5

Materializing Postenvironmentalism in Living Spaces

Abstract How does a material-semiotic postenvironmentalism actually take form in the world? This chapter presents some examples (including the Transition Network movement, the U'wa's and the Brazilian seringueros' struggle...) of current environmentalist practices that while confirming environmentalism is not dead at all, nonetheless show it is transforming by including nonhuman and more-than-human networks in the realm of social actors and by listening unheard voices through devices, techniques, and procedures that allow their expression in the public space. Their space of interaction is a "living space", which is the locus for environmental issues to be pragmatically debated in forms of life. The three-step process of assembling, mobilizing, and impacting, which characterizes environmental actor networks' agency, is finally considered.

Keywords Material semiotics · actor-network theory · living space · urban smartness · urban gardening

In 2006, when the post-environmentalists' message entered public debate by criticizing the strategies of both the institutional and the grassroots movements, the *Transition Network* started its first activities in the village of Totnes, Devonshire, UK. Over the course of a year, its founders, Rob Hopkins and Naresh Giangrande, organized a large

© The Author(s) 2016
C. Certomà, *Postenvironmentalism*,
DOI 10.1057/978-1-137-50790-7_5

number of meetings addressing the local community and designed to raise awareness about climate change and peak oil. The Transition Network's model rapidly attracted the attention of a large number of scholars and citizens, spreading their relocalization efforts around the globe and making the model replicable elsewhere. The network inspired and supported local people in their resilient community-building processes and in drastically reducing CO_2 emissions, including initiatives such as sustainable housing, domestic energy saving, creation of a local currency, and local food markets. Private founders supported the emerging movements, and courses were delivered in Australia, Canada, the UK, Germany, Ireland, Italy, the Netherlands, New Zealand, South Africa, Spain, Sweden, and the United States. The *Transition Network* initiatives work as context-dependent, do-it-yourself, and technology-equipped affirmative actions advancing the "power of just doing stuff" (Hopkins 2013) by assembling a number of local actors (including local governments, energy suppliers, building companies, and local retailers) in the work of observing, mobilizing, and remaking the spaces in which they live—and, in so doing, make it possible for them to have an impact on a larger scale. As a spontaneous response to the failure of the Copenhagen Climate Change Conference in 2010, as well as to the geopolitical and economic threats deriving from fossil fuel depletion and the reality of global warming, the *Transition Network* claimed that grassroots projects can effectively engage the local community to make a real change in one's hometown, and that this can happen everywhere. The upsurge of local, spontaneous, awareness-raising, and sustainability-oriented processes and practices suggest that we are not witnessing the definitive end of environmental thinking, but rather experiencing its transformation into something new.

Maybe all is not lost: there actually exist promising alternatives to merely resignedly accepting the progressive fading of environmentalists' inspiring message. Postenvironmentalism is already transforming the death of environmentalism into a renewal of inspirational practices by mobilizing worldwide, forward-looking, and heterogeneous energies. A number of grassroots, local-based initiatives have flourished in large portions of the world in the recent years[1] and provide evidence of the current transformation of environmental thinking. Most often, they do not claim for themselves the name of postenvironmentalism;

nonetheless, their features can be easily ascribed to the particular form of environmental thinking and practice described in Chapter 4. All of these cases blur the boundaries between real and virtual, natural and artificial, and human and nonhuman and make evident the relevance of a paradigm shift toward a form of network thinking whose aim is to change the world starting from material practices in local spaces and thus advance innovative perspectives. While it is clear that postenvironmentalism does not provide off-the-shelf solutions for saving the world from the incumbent environmental crisis, it is probably also clear that this is not its purpose anyway. Rather, by debunking mainstream environmental thinking (see Chapter 2) and the (merely) critical account of different post-environmentalisms (see Chapter 3), it aims at offering an alternative way of thinking (through) the world—one which may eventually change it (see Table 3.1).

Examples in the following paragraphs explain how the ontological, epistemological, and political dimensions are addressed by material-semiotic postenvironmentalism. These outline how humans, nonhuman, and more-than-humans express the public debate by adopting a plurality of (linguistic and nonlinguistic) means, which include opposing resistance or practicing resilience, refusing the roles prescribed to them, or producing previously unthinkable social formations. In addition, the examples show how these means determine different spatializations. The heterogeneous networks begin by uprooting their epistemological basis and restructuring our relationship with the environment (e.g., breaking down the traditional distinctions of natural/technological, real/virtual, human/nature) and perform postenvironmentalist practices in everyday life and space. This might take the form of an antinomic re-creation of nature in the city and the mobilization of nonhuman companions in advancing a bodily form of political protest or the transformation of urban material infrastructures by using the capability of the social web to bring together collective knowledge.

5.1   Space Matter(s): On the Emergence of Living Spaces

Material semiotics' interest in the matter and spatiality of everyday life practices has drawn interest and inspired a plethora of research and initiatives aimed at unveiling and disarticulating the hegemonic structures of power distribution, and thus advanced an openly inclusive understanding of global relationships that is sensitive to inequalities. This point particularly resonates with political ecology approaches and substantially affects the definition and treatment of environmental issues in postenvironmentalism. Thanks to the lessons of political ecology, postenvironmentalism does not conceive social issues as distinct from ecological ones, and while this complexifies the story, it also offers a more complete understanding of the roots, causes, and possible solutions of these problems. In addition to the strong focus on the material constitution of environmental issues, it also provides a particular understanding of the space of environmental negotiation as a "living space" generated by the continuous work of assembling purpose-oriented networks around a practical matter of concern (rather than around ideological claims). In fact, networks gather together diverse actors around a matter of dispute: ecologists with their narratives, symbols, and claims; academic scholars with their research procedures and devices; citizens with their normative tools; animals with their communication practices; and even territories and their specific chemical composition. Their mobilization necessarily entails a spatial dimension and generates a vibrant context where matters of concern are practically addressed. Agreement or disagreement among actors, as well as different opinions about the fate of an issue, always arises in the form of material configurations of beings, things, and processes and are thus inseparable from specific contexts, precisely because actors are kept together and made able to interact on the basis of their sharing a common space of life (see Fig. 5.1). The space defined by postenvironmentalist practices has multiple functions and is constantly made and remade by complex assemblages mutually affecting each other. It is not only actively produced by administration, businesses, and citizens, but it is also "claimed" by the sea, invaded by the birds, transformed by factories, flooded by rains, connected by internet cables, etc. All of these things exert their agency on space by imposing/proposing a particular view of what a place should be and showing a possible way of "dwelling" it. Space itself is thus an event created by the material overlapping of heterogeneous assemblages gathering around disputed issues; it is only momentarily present and can be dispersed again; it is mutable, radically open and unstable, located at the crossroad of power and knowledge relations.

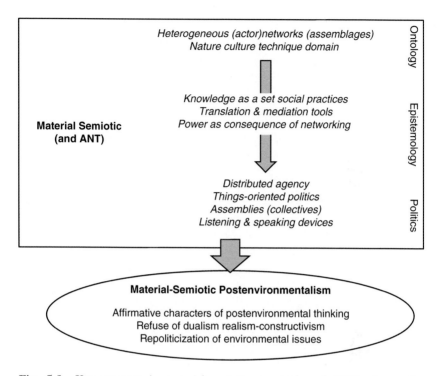

Fig. 5.1 Key concepts in material-semiotic approach and ANT relevant for postenvironmentalism

The living space of postenvironmentalism is thus at once (a) a material-semiotic agglomerate generated by the gathering of multiple actors around a disputed state of things; (b) the locus for environmental issues to be conceptualized and addressed by collectives; and (c) an issue in itself for the public to be concerned about.

Box 5.1 Living spaces
(a) Living space as an effect of actor-network assembling
As already mentioned, postenvironmentalism first and foremost offers a new understanding of our relationships with the nonhuman and more-than-human world, not so much in abstract terms as by disclosing the

pervasiveness of hybrid formations and heterogeneous linkages in the spatial texture of our daily life. A clear example is provided by geographers Steve Hinchliffe and Sarah Whatmore in their (postenvironmentalist) exploration of Birmingham's "recombinant ecology" which makes the city livable for different kind of inhabitants, as well as a source of identity and association for them (Hinchliffe and Whatmore 2006). Hinchliffe and Whatmore claim that spatial division between civic and wild, town and country, human and nonhuman is nonsense in the urban context; cities are heterogeneous formations emerging from the interaction between multiple and disparate actors. The case of Birmingham exemplifies this. Birds, particularly peregrine falcons, have made the city their habitat; even if they were not directly encouraged to do so, the falcons' choice of Birmingham as site for settlement is probably motivated by the abundance of food (such as pigeons unaccustomed to predation), and the warm and hospitable microclimate of the city. In Birmingham, peregrine falcons settled in the city center, under satellite dishes of the tower (a place theories of urban planning envisaged as specifically human), and their presence, together with pigeons and humans, contribute to the design of the urban space. Not far from the city center, in an artificial water basin, 10 species of fish and otters (which enter via urban canals) have been recorded. Black redstarts—Britain's rarest bird—have made their breeding grounds in a nearby abandoned building. The list could be continued. Hinchliffe and Whatmore thus ask: **Who** is making the city? How is the cityscape forged? Their findings show that nonhumans are active agents in forging urban places, and, by converse, their life is constantly influenced by urban relations—even though their ecology might be completely different from the one observed in wild contexts. The result is a peculiar living space where ecologies become urban and cities become ecological (Hinchliffe and Whatmore 2006).

(b) Living space as a locus for the collective to debate environmental issues

Ecological issues gather a broad and diverse range of actors around common matters of concern and offer them the opportunity to restructure themselves as hybrids of nature and culture. In so doing, they activate cross-scalar spatial connections and mobilize flows of matters and meanings by producing a distinct space for the general public to turn into an issue-oriented collective. Such a space is characterized by a network structures in which relational proximity—rather than geographical proximity—plays a key role in generating action and in making evident the existence of narratives and imaginaries that are mobilized in the debate around a disputed state of affairs. For instance, in 2001 the indigenous people U'wa, living in the Colombian territory and amounting today to about 7,000 individuals, threatened to commit collective suicide by jumping off a sacred cliff to protest against oil exploration activities on their land by the US *Occidental Petroleum* Corporation. The space of the oil extraction issue crossed geographical borders and generated a parliament of actors mobilizing around the world. The articulation of the debate most often took an indirect and practical form, so that the exploitation process was directly contested through bodily occupation of land (including roadblocks and sit-ins at the drill locations), and also with more impact through international environmental campaigns (most notably by the associations Amazon Watch and the Rainforest Action Network) which made full use of all available communication methods (including internet communication). This overlaps with the narratives about rainforest and the oil described and made the object of traditional rituals as the blood of the Earth by the Werjayà, or spiritual leaders, in the U'wa community. National laws on rainforest preservation and international human rights regulations entwined with the evidence of the U'wa title to the land dating back to 1661. All these combined with the $1.5 billion sent from United States to Colombia to support military forces in the region and the counter-flow of funds collected from around the world allow U'wa to buy their land (Van Haren 2015; Cultural survival 2015). The living space of U'wa vs Oxy spanned from a few acres along the Colombian-Venezuelan border to the Occidental Petroleum headquarters in Los Angeles which grassroots environmentalist groups occupied, to the Bogotà bidonvilles where some indigenous people had already displaced, and finally to the international routes of U'wa representatives in Europe and North America as they traveled to gather support for their cause. Just a few acres of forest became a globally contested issue and gathered together a complex network linking together heterogeneous actors, bringing their own material practices, interpretive systems, power geometries, and knowledge and places (Escobar 1999). The space of the U'wa claim against oil extraction emerges thus as a living space where postenvironmentalist thinking articulates its resources and power to challenge the status quo and, from a tiny spot of forest, had an effect on the globe as a whole.

(c) Living space as an issue for postenvironmentalism

Living space is at once both the context and the object of postenvironmentalist issues. It is important to be clear that, unlike the nostalgic interpretation of natural places as bulwarks against the postmodern disenchantment of the world (Certomà, 2009), here they are regarded as complex and dynamic assemblages able to attract the interest of different actors. Postmodern philosopher Donna Haraway's reading of the story of the Brazilian seringuero Chico Mendes highlights how network configurations produce a living space that is the location for and at the same time the object of a political ecology struggle (Haraway 1992). The union of the extractors and the indigenous people led by Mendes derived their true position as defenders of the forest from their daily relation with the forest and from their claim that forest economy and management was an integral part of their struggle to survive. The seringueros' narrative makes it possible to deconstruct the image of the tropical rain forest as Eden under glass and advances a view of environmental issues as social issues, rather than as just saving nature. As Haraway suggests, the very novelty and power of Mendes' fight did not derive from a representation of nature as something distant and distinct from the human domain, but in addressing the constitutive relational character of forest-dwellers and the possibility for the forest to engage a number of actors in defining its destiny (Haraway 1992). Living spaces can thus be regarded as mobile and fuzzy events whose features are continuously—still in some case, imperceptibly—mutable. The tropical forest, for instance, according to Haraway, is not a physical place, not a treasure to fence, not a code to be read using mathematical formulae or scientific models. It is "neither mother, nurse, nor slave, ... is not matrix, resource, or tool for reproduction of man [; nature] is, strictly, a common place ... widely shared, inescapably local, worldly, enspirited ..., is the place to rebuild a public culture" (Haraway 1992, 296). This view recognizes the centrality of space in both hosting and inspiring the conjunction of human and nonhuman trajectories.

Environmental thinking performed in local places using material-semiotic postenvironmentalist practices is actually already providing an alternative to the death of environmentalism. As planning scholars Damian White and Chris Wilbert have described, there are a number of activist initiatives giving shape to this new trend in environmental thinking:

> Consider the attempts by the Bioneers in California to draw together currents of Green activism with progressive engineering; or Brian Milani's eco-materials project in Toronto which seeks to draw together trade unionists with environmental activists and engineers to explore communalist and democratic expropriations of the diverse possibilities of the 'new productive forces' from industrial ecology and post-Fordist ecological technology. [...] All these interventions indicate an increasing desire by many activist as well as academic currents to move environmental debates beyond stale dualisms and oppositions, such as that between technophobia and technophilia. In some senses, there is a structure of feeling in these 'cyborg ecologies' that there is now no going back to any kind of purism of the natural. (White and Wilbert 2006, 101)

All of them are attempts not to preserve the purity of nature, but rather to recover the joy of being green and its sense of possibility. They suggest that rather than complaining about the "malaise of modernity" (Taylor 1991), we can become involved in socio-ecological and socio-technological processes that are able to generate living spaces of environmentalist engagement (Hinchliffe 2007).

5.2   ASSEMBLE, MOBILIZE, IMPACT!

One of the most relevant critiques to post-environmentalism has been formulated by Blühdorn: it points out the lack of identity and thus its inability to inspire commitment to the cause. However, I argue that the material-semiotic perspective can lead us to define a post-environmentalism endowed with its own distinct characteristics (which I named postenvironmentalism), which does not merely add further values and rules to the exiting environmental theories. Postenvironmentalism is necessary to subvert our very understanding of what environmental thinking is intended for, and what issues it is actually about. This affirmative approach builds upon the recognition of the role of materiality and space in the formation of heterogeneous assemblages that define in common the fate of an issue on the basis of common and practical concerns.

Three steps characterize postenvironmentalist initiatives: the assemblage of actors, their mobilization, and their effect on the state of things.

The first step requires actors to gather around an issue they are concerned with; most frequently in the case of environmental issues this relates to the conditions and possibilities for the actors to dwell in a particular environment. While this might recall a constructivism position by suggesting that environmental issues do not contain in themselves the reasons for their importance as they are a mere construction of environmentalists, postenvironmentalism actually maintains something radically different. It claims that environmental issues are common constructions of humans, nonhumans, and more-than-human actors that gather around a real matter of fact and turn it into a matter of concern in the public debate. Gathering a crowd, generating an assemblage, forming a network are all different ways of expressing the spontaneous linking process that puts in close relationship dispersed things/beings and events around a disputed affair that they all—in varying degrees—have an interest in (see Fig. 5.2).

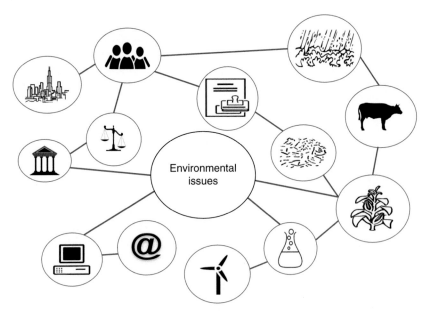

Fig. 5.2 A simplified schematization of issue-based actor network

Source: the author

The assembling step generates a network of actors brought together, via a sort of coalescence, by their very condition of being all affected by the issue they gather around; the network thus springs in a quasi-deterministic sense from a certain configuration of the world. The form and strength of these linkages as well as the process by which they are established is variable and conditioned by the peculiar characters of the involved actors (Barry 2001; Bennet 2010). While it is evident that there are some actors endowed with greater capability to assert agency, such as multinational companies, it is perhaps less evident that there are some nonhuman actors, such as climate (especially in the form of changing climate), that have a great aggregating potential compared to others. In most of the cases assembling around a matter of concern and networking in order to produce significant and potentially impacting linkages is a spontaneous process that does not entail central coordination.

The interaction between involved actors is determined by their own willingness, constitution, and status; this is at the origin of the mobilization process. It is activated by a material-semiotic shift that modifies both the matter and the significance of the context, while making some actors able to gain a catalytic role and working as attractors of others. Focal agents create and maintain new loyalty linkages with the aim of changing the state of things toward new configurations. In the emerging parliament of things (laboratories, plants, markets, streets, etc.) where public issues are constantly addressed, a sort of material-semiotic discussion process is activated by the presence of mediation devices that make mute things speak and allow speaking things to listen. Community energy project networks are an example of networks assembling around the issue of saving and generating energy; as *The Rough Guide to Community Energy* (Clark and Chadwick 2011) suggests "The beauty of these projects is that they can help deal with so many problems at once, making a difference at the local and global levels simultaneously" (Clark and Chadwick 2011, 7). Unlike mainstream energy policy approaches, this brings to the fore a number of interconnected actors (which can, under certain circumstances, also play the role of issue in themselves, e.g., wind turbines), rather than calling for the creation of global regulations to solve the issue. From being mere customers and energy users, citizens here turn into innovators and energy producers, their potential to disclose knowledge and their role as active solution seekers turn them into focal agents able to assemble a complex network, including the national energy grids, the regulation for energy production and distribution, existing infrastructures, the configurations of home and private space, loan provision services, migratory birds, noise inspection devices, and many others.

The nature and function of technologies in postenvironmentalism is a tricky point. For instance, the revolutionary role of information and communication technologies (ICTs) has been recognized by popular media, where they are frequently celebrated as the key to a new "environmentalism 2.0." Nonetheless, this common understanding simply describes the tip of the iceberg, but it does not mention the real revolution in terms of identity creation, knowledge production, and political governance that ICTs bring forth. In fact, Internet technologies will not save the world, as they are *already* part of the world as it is, but they can increase relationships and connectedness—and thus they can support actions that have an impact on the current state of things. Particularly, in terms of power relations, ICT-based networks generate transactional governance that includes temporary, purpose-oriented, and dynamic alliances. Regardless of their belongingness, status, or ideological stances, different actors may take part in unexpected coalitions and share responsibilities on the basis of a purpose-oriented agenda whose priorities are constantly negotiated. In addition to acting as actors themselves, technologies such as sensors, the Internet of Things, and social media are now able to equip actors and fuel new forms of collective mobilization by expanding the meaning of environmental commitment. In this regard, the role of spontaneity in the creation of social formations has been recently pointed out by a large number of scholarly contributes, as we are becoming more attentive and listening to the way heterogeneous actors create an impact on the world, rather than the way we can master and induce them to do so.

Box 5.2   Spontaneous networking for urban smartness
In recent decades, cities have emerged as a space suitable for experimenting with innovative strategies to advance sustainable development innovations (Istanbul Declaration and Habitat Agenda 1996; IPCC 2007; UN-HABITAT 2010). A number of appealing definitions, such as resilient, transient, smart cities, have been introduced to characterize the sustainable city of the future. "Smart city" in particular rapidly turned into a buzzword that has been used with reference to almost any technology-driven urban initiative, encompassing a broad

range of aspects of urban life (including quality of life and welfare, sustainability, social cohesion, and economic growth). What differentiates it from already-existing projects that merge urban planning, social engineering, and innovation is the idea that technologies will play a crucial role in urban life and solve our current environmental problems without significant changes in our way of life (March and Ribera 2014). However, the senselessness of many smart city initiatives—ranging from the construction of new completely planned cities characterized by the massive presence of technologies of control and organization, to disconnected initiatives of local food promotion—have exposed the concept to a wide range of criticism (Vanolo 2014). These particularly exposed the "regimes of accumulation triggered by new technologies; the role of major multinational enterprises in shaping technological imaginaries; the mobile political technology strongly connected to neoliberal rationales" (Ribera, Santangelo, and Vanolo 2015). The interest in the technical improvement of city infrastructure and technological application of data-driven solutions often disregards the evidence that although these are important, they are not endpoints in themselves.

If we consider for instance the way in which sustainability governance processes are today conducted in cities, it is clear they do not emerge (or only partially do so) from top-down regulation but as creative effects of encounters between heterogeneous actors. Collaborative urbanism initiatives mapped by the Spontaneous Intervention network or the crowdsourcing-based *(Im)possibleLiving* project to restore derelict buildings exemplifies how the spontaneous bottom-up agency of socio-technological networks produces non-planned smart processes.[2]

It is thus possible to advance an alternative understanding of "actually existing smart cities" (Shelton, Zook, and Wiig 2015) not as the outcomes of top-down government programs or the business strategy of major technology companies (Söderström et al. 2014), but rather as emergent effects of autonomous and self-empowering practices performed by network-shaped social actors equipped with dedicated speaking and listening devices. This understanding of smartness recognizes the crucial role of technology in the process of producing collective strategies and enabling people and nonhuman actors (e.g., the Internet of Things and the integrated environmental sensors networks) to take part in the public debate (Latour 2004). If we consider, for instance, the diffusion of crowdsourcing processes (including citizens' science, participatory sensing, social mapping, and computational thinking) and tools (e.g., smartphone software, blogs, wikis, social bookmarking applications, social networking, peer to peer networks), it becomes clear how ICTs offer manifold possibilities for transforming science production practices by harvesting crowd-generated data and knowledge and making them available via open access databases and free software (Brabham 2009, 2012). Crowdsourcing enables a number of web-based problem-solving processes by involving large groups of users that can perform functions difficult to automate or expensive to implement. It has been recently recognized by scholars, planners, administrators, businesses, and civil society as an appropriate process to support the emergence of a new governance model characterized by distributed technological agency in the transition toward a new urban-based environmental thinking (e. g., the *EveryAware project, Mapping for Change Citizen Cyberlab, openIDEO, Extreme Citizen Science* groups).[3] By equipping the actor network with dedicated software and devices, crowdsourcing fuels the possibility for humans, nonhumans, and more-than-humans to bring forth their "opinions" on a debated issue in a common parliament where traditional hierarchies lose their supremacy, as well as to contribute to both knowledge production and decision-making processes. Evidence indicates that crowdsourcing platforms are increasingly tackling socio-environmental issues (Certomà, Rizzi, and Corsini 2015), and in this regard, they can foster far-reaching genus of heterogeneous actor-network users who are shaping the future of urban sustainability processes. Similar initiatives can be regarded as postenvironmentalist exactly because they mobilize technological devices and processes by linking together heterogeneous social actors that negotiate in common a broad range of sustainability-related issues and are able to generate smart solutions based on coordination and mutual support (Sennet 2012).

In postenvironmentalism, debate and negotiation about environmental issues may well involve words, but they are not discursive in kind, as different agents may have their say by using other forms of expression: the declarations of international meetings, union strikes, geological resilience, machines working or not working under specific weather conditions, administrative resistance to rules changing, animal species proliferating or becoming extinct, alterations in the chemical composition of water, parasites in crops, etc. Politics is not the results of the discussion between alternative ideologies and models competing in the traditional parliamentary arena; rather it is the very act of trimming relationships with other actors assembling around a common matter of concern and aligning their behavior toward a certain state of the world.

Box 5.3 Postenvironmentalist gardening

Urban gardening emerged as a movement aimed at engaging people in gardening in the city (whether growing food, tending land, planting flowers and trees, or breeding animals) as a way to address crucial socio-environmental issues such as the scarcity or poor quality of public spaces, the lack of green infrastructure, the need for better human relationships with nature, and the urgency of providing marginalized social groups with dedicated spaces for self-improvement. A classification of the entire panorama of urban gardening activities is almost impossible to provide, because different countries, traditions, and contexts generate vastly different gardening practices; however, the most common categories include allotment gardens (i.e., portions of public land provided, planned, designed, and regulated by the local authorities); community gardens (i.e., areas of public or abandoned private land where citizens run gardening projects aimed at community building) (see Fig. 5.3); and guerrilla gardening (i.e., flash mob-kind actions intended as political gesture to bring attention to the need for more green and accessible areas). In general, these activities are regarded as positively influencing the environmental and social quality of city space and people's life as they intended to meet such aim as education, leisure, and socialization (Wekerle et al. 2009); counteracting food insecurity (McClintock 2008; Pinkerton and Hopkins 2009; Milbourne 2012) and social disadvantages (Emmett 2011; Schmelzkopf 1995); community building (Beckie and Bogdan 2010; Been and Voicu 2006) and health promotion (Wakefield et al. 2007); involving marginalized social groups (Tracey 2007; Flachs 2010); and finally advancing an environmental commitment (Miller 2005; Certomà 2011). Building

Fig. 5.3 A view on "Orti Urbani Garbatella", one of the first urban gardening initiatives in Rome

Source: Zappata Romana

upon shared ideals, gardeners often establish links with other informal planning initiatives, including alternative economic networks (Kurtz 2001), transient cities or urban green renovation programs (Pagano and Bowman 2000), or projects for accessibility of disadvantaged people (Ferris et al. 2001).[4]

Urban gardening envisages a creative intervention into the spatiality of daily life (Hou et al. 2009) emerging from people's social and political engagement in the transformation of the material and symbolic constitution of the city (Schmelzkopf 1995). As activist gardener David Tracey explains:

> The reason why you're doing this … is not just to make one patch of the neighbourhood more pleasant to look at, but because you believe the entire city is worth the effort. And because you decided that,

rather than wait for the world you want … to just appear, it was better
to start making it yourself. (Tracey 2007, 15)

Gardening initiatives entail a number of material-semiotic practices
that can be easily understood as postenvironmentalist because they
do not focus on theories but on practices brought about by a
collective mobilization of human and nonhuman actors in the city
space. They thrive on the differences of styles, plants, attitudes,
tactics, and traditions that are needed to address environmental
issues directly and immediately in the place of life. Biological
material is used as a means of political expression when urban
gardeners forge alliances with plants, animals, the built environment, the weather conditions, and the urban ecosystems. While
urban gardening is an immediate and simple way to arrange city
streets, plots, squares, or flowerbeds, gardeners are ultimately
motivated by global political issues (such the privatization of
space, international and national redistribution of land, north-
south trade relationships, green issues and environmental risks,
the general decrease in the welfare state, and the problem of food
security).

The connection with the place and matter of daily life makes
evident forgotten or ignored ecological relations, which turn into a
public issue through the association with nonhumans and more-than-
humans in the co-constitution of urban space. Urban gardeners'
understanding of the ecological reality opens the door to an alternative form of political representation of ecological claims in the
cities; and, it requires a collaborative relationship with living green
spaces as place of interest, re-appropriation, and care. At the same
time, this induces a re-politicization of environmental thinking
because

> [e]very bit of land you see around you, from the lawn across the
> street to the street itself to the schoolyard at the end, is used accord-
> ing to a decision made by someone. The decision may not have
> involved you at the time, but you're involved now because it makes
> a difference in kind of the world you live in and react to every day. If
> land matters, so too do all the things that may or may not grow on it.
> (Tracey 2007, 32)

NOTES

1. This includes, for instance, worldwide fixers of the *iFixit* community fixing the world "one device at a time" (information available at https://www.ifixit. com/); the *Permaculture Network* of desert greeners (information available at http://permacultureglobal.org/); or the resilient community planners (information available at http://thrivingresilience.org/trcc-overview/).
2. Information available at http://www.spontaneousinterventions.org/ and http://www.impossibleliving.com/
3. Information available at http://www.everyaware.eu/; http://mappingforch ange.org.uk/; http://citizencyberlab.eu/; https://openideo.com; https:// www.ucl.ac.uk/excites
4. A repository of European initiatives has been provided by the research network on "Urban Allotment Gardens in European Cities – Future, Challenges and Lessons Learned" founded under the COST framework (http://www. urbanallotments.eu/case-studies.html); a further source of information about worldwide initiatives is the Guerrilla Gardening website, which reports actions nearly every continent (http://www.guerrillagardening.org/).

CHAPTER 6

Summary and Conclusion

Abstract We are now at the conclusion of this journey through the development of environmental thinking, which explored the debate on the mainstreaming and subsequent death of environmentalism, together with the most recent scholarly and practitioners' contributions on post-environmentalism. This chapter, thus, summarizes it and, building upon the transformation of post-environmentalism into postenvironmentalism (a transformation entailing a new worldview, rather than the simple removing a hyphen), it traces a tentative manifesto of the future environmental thinking in action. Particularly, it criticizes the idea that we are really facing the end of environmentalism and, rather, suggest that new postenvironmentalist practices are already mushrooming worldwide.

Keywords De-politicization · environmentalism institutionalization · material semiotics · postenvironmental agency

In Chapter 1 of this book, we considered how, over the last few decades, environmental issues gained a central role among other political concerns; they entered academic departments and have been listed among the most worrisome contemporary emergencies by citizens around the world. Environmentalist commitment took a number of diverse forms, from international environmental diplomacy, leading toward the establishment of a global environmental regime under the auspices of the UN up to

© The Author(s) 2016
C. Certomà, *Postenvironmentalism*,
DOI 10.1057/978-1-137-50790-7_6

and including a manifold of grassroots organizations that proposed alternative environmental friendly lifestyles (such as the radical environmentalist movement *Earth First!*, or the *Global Ecovillages Network*).

At the same time, the media has frequently reported environmental problems, and related concerns have been included in many companies' corporate social responsibility reports. For instance, many of the biggest multinational oil companies, charged by environmentalists of causing irreversible environmental pollution (e.g., the British Petroleum's oil spill in the Gulf of Mexico, 2010) and of violating human rights in regard to their relationship from the environment (e.g., the environmental damage charges put forward by the Ogoni people in Nigeria against oil companies Shell, Chevron Texaco, and Nigerian National Petroleum Company), have now included environmental responsibility in the corporate policy.

This process of mainstreaming environmental values in the context of liberal democracies followed two principal directions: the realist trend, which claims that solutions to environmental problems can be provided by the advancement of scientific knowledge and technical measures (most notably regulatory norms or market-based instruments; e.g., the tools for managing CO_2 emissions market defined by the Copenhagen Climate Change Conference in 2009); and the constructivist trend, based on the idea that appropriate solutions to environmental problems require a radical change in cultural attitudes and behaviors (e.g., the *Agenda XXI* provisions provided by UN Summit in Rio de Janeiro in 1992).

Despite joint efforts by international institutions and large environmental organizations to move toward consensual environmental politics, the results have not been impressive, and the target goals defined in official declarations have been generally disregarded. In his text on the future of environmentalism, Castree describes this apparently paradoxical situation. He suggests that after the UN Earth Summit, held in Rio de Janeiro in 1992, "green thinkers" assumed that their proposals had become part of everyday common sense, a belief that resulted in a fatal misunderstanding: what had gained a strategic foothold in the international political arena was not environmentalism in general, but one specific form of it, namely the eco-liberal orthodoxy. This apparent success actually hid a critical failure of the whole environmental movement, because while environmentalism appeared to exert real societal influence, in practice it was mostly ineffectual. To a large extent, environmental discourse has been appropriated by a large number of actors with very few green credentials, while "environment" became a global buzzword. At the same time, although the public generally indicated that the environment was something it cared about, this did not necessarily translate into concrete actions. Therefore, Castree (2006, 14) affirms, "It is plausible … to suggest that Western environmentalism today has many supporters but little purchase, lots of popularity but little power, many advocates but few serious practitioners."

6.1 Against Environmental Thinking Institutionalization

Starting in the 1980s, international environmental organizations and large NGOs began a process of de-politicization of environmental issues that reduced the influence of environmental thinking and related political initiatives. The exclusive focus on technical or procedural solutions to environmental problems weakened the political, social, and cultural strength of the early environmentalists' claims and destroyed social agency so that general citizen efforts (such as reducing waste, buying organic, using public or green transportation and recycling) were frustrated by the institutional inability to implement environmental protection measures, resulting in an even further loss of credibility for environmental thinking.

Despite widespread idea that external causes determined the failure of environmental initiatives (such as laziness by the general public or polluting companies' unwillingness to comply with green principles), there are reasons to believe that internal reasons played an equally significant role in making environmentalism ineffective. These pertain to the very nature, objectives, and means of environmentalism itself (Darier 1999; Katz 1998).

Most notably, a public debate arose in US society when Schellenberger and Nordhaus, environmental consultants at the Breakthrough Institute, published the pamphlet *The Death of Environmentalism: Global Warming Politics in a Post-Environmental World* (2004)—a debate that soon spread across the Atlantic. However, in Europe "post-environmentalism" was not new; in the wake of postmodern theory, John Young had already published a book titled *Post-Environmentalism* (1992). Young's work describes the failure of the radical environmentalism of the 1970s, which was said to be excessively concerned about nature and not enough about society, and calls for a new and effective environmentalism able to address the pressing challenges associated with the rise of poverty and inequality in the age of globalization. The word was soon appropriated by other Critical Theorists, i.e., those scholars that elaborated the critique of late modernity on the basis of the legacy of Frankfurt School and particularly on Jurgen Habermas' theory of communicative action. They characterized post-environmentalism as a further step toward raising public awareness of the relevance of semantic and discursive practices in the construction of environmental issues.

Still, it was not until the publication of Schellenberger and Nordhaus' pamphlet that post-environmentalism reached the general public, first in the discussions of activists, then reaching the academic community and generating an increasing number of publications in specialized journals (most notably the review *Environmental Politics*). This produced a sort of post-environmentalism craze, paradoxically resulting with many of the oldest established environmentalist associations claiming to be post-environmentalist (e.g., the conservationist Italian association *FAI—Fondo per l'ambiente italiano* (Fund for the Italian Environment); Neri 2004).

Schellenberger and Nordhaus' study is important not just because it succeeded in popularizing the term post-environmentalism, but because it points out the principal reasons that environmentalism and mainstream environmental politics lost much of their attractiveness. Their research focuses on how environmental issues are presented reductively as sectorial interests that broaden the gap between social and environmental domains, rather than bridging them. They pose the rhetorical (but often neglected) question: "Why, for instance, is a human-made phenomenon like global warming—which may kill hundreds of millions of human beings over the next century—considered 'environmental'? Why are poverty and war not considered environmental problems while global warming is? What are the implications of framing global warming as an environment problem—and handing off the responsibility for dealing with it to 'environmentalists'?" (Schellenberger and Nordhaus 2004, 12)

The interrelatedness of social and environmental phenomena is today largely acknowledged and analyzed by political ecology scholars who elaborated the theory of environmental justice and environmental conflicts (Martinez Alier 2003). They point out, for instance, the mutual influence of social conflicts and environmental pressures (Homer Dixon 2001), the pressure of climate change on international migrations (El-Hinnawi 1985), or the self-reinforcing cycle of environmental degradation and economic, social, and cultural poverty (Soja 2010).

However, Shellenberger and Nordhaus' question, while rhetorical, is not trivial. In fact, although the interrelatedness of social and environmental phenomena is common knowledge for environmental experts and activists (it was, for instance, already described in 1994 in Ksentini's report *Human Rights and the Environment* (Ksentini, 1994)), it may not be self-evident for the general public. This implies that in some cases, it is very difficult to arrive at agreement about political priorities or the allocation of funds when competing issues require public attention. Because the complexity of socio-environmental phenomena makes it difficult to understand and address them appropriately, mainstream environmentalism frequently adopted a strategy of simplification and presented its concerns in rather naïve terms. This means that the roots of a given environmental problem is identified as having a single cause (which can ideally be addressed by a specific technical solution), and no further investigations are conducted to detect deeper reasons or hidden connections for its occurrence. Shellenberger and Nordhaus explain, for instance, that many Americans believe the causes of every environmental problem can be addressed by a single policy measure (as it was the case for the acid rain problem that the American public considered solved by the enactment of the 1990 *Clean Air Act* regulating airborne contaminants). However, most environmental problems require a careful analysis of other-than-ecological aspects, as they may have deeper or remoter causes in social, economic, cultural, or political conditions and also require large alliances (including environmental experts, environmentalists, social scientists, technicians, social organizations, etc.) in order to develop and implement effective and lasting solutions. Forging such alliances, however, is very challenging, as the debate between environmentalists and anthropologists confirms.

From the 1980s onward, anthropologists and cultural studies scholars raised criticisms of environmentalists' exclusive interest in the natural environment and called for a greater attention to be devoted to the cultural side of environmental issues. This constructivist approach strongly influenced the Critical Thinkers' understanding of post-environmentalism. Constructivists affirm that what counts as reality depends on the observer's perspective, and there is no easy way to separate objective observation from social biases. Not only the foundation of knowledge (i.e., the social heritage individuals acquire in their becoming part of a community and through experiential learning), but also the foundation of science (i.e., a systematic, objective, and methodologically derived body of explanations and predictions in specific fields of knowledge) is, actually, the result of semantic practices. This means, for example, that the very nature investigated by ecologists is, itself, a cultural artifact. For instance, anthropologist

Kay Milton summarized what happens when personally confronted with socio-environmental problems in the field:

> The environmentalist in me wants to get on with the work, to plant trees, lobby politicians, stop pollution, save the whales and the woodlands, halt the destruction wrought by the blind pursuit of profit and "progress." The trained anthropologist, irritatingly, wants to stop and ask questions. Why do we believe what the scientists tell us? Why do we consider whales and woodlands important? What kinds of assumptions underlie the claim that the Earth is in danger? (Milton 1996, 2)

Milton's observation reflects the prominence of the realist approach adopted by global environmentalism with the support of environmental institutions.

The emergence of scientific environmentalism in the 1980s and its inclusion in the international diplomacy arena were in fact accompanied by a chorus of critiques, principally advanced by cultural studies scholars and political ecologists and also by deep ecologists, eco-anarchists, post-colonialist thinkers, and activists interpreting environmental commitment as contestation and resistance to economy-driven modern culture (Taylor 2013). They questioned the reliability of scientific results, methods, and suggestions, and whether it makes sense to lend natural scientists the power to define the global political agenda for the years to come (jointly with politicians and business lobbies). Most of them pointed out the very issue that is today at the /core of the post-environmentalist critique, described by sociologist Wolfgang Sachs in the following terms:

> scientific environmentalism construct[s] a reality that contains mountains of data, but no people. The data do not explain why Tuaregs are driven to exhaust water holes ...; they do not point out who owns the timber shipped from the Amazon or which industry flourishes because of a polluted Mediterranean sea; and they are mute about the significance of forest trees for Indian tribals ... In short, they provide a knowledge which is faceless and placeless ... It offers data, but no context. (Sachs, 1993, p.22)

6.2   ARE WE REALLY WALKING ON THE EDGE?

The second half of this book focused on the edge of environmental debate, where different conceptions of post-environmentalism compete for public attention. On one side, Schellenberger and Nordhaus' post-environmentalism proposes reinvigorating political emotion for development under green auspices in the context of the neoliberal economy. In this sense, the definition of post-environmentalist is closely resonant with the optimism of ecological modernization theory (Christoff 1996; Lomborg 2001), a strategy described by Nobel Prize winner and former US Vice President Al Gore in the reportage *An Inconvenient Truth* (2006). Shellenberger and Nordhaus' concern with redevelopment requires emancipation from nature rather than closer engagement (Latour 2008) and thus further deepens the problematic distinction between social and environmental issues.

On the other side, cultural scholars describe the future of post-environmentalism by introducing the idea of "post-ecologism", which digs deeply into the semantic implications of the discursive production of nature and ecologies (Blühdorn and Welsh 2008). Their progressive attempt at overcoming the theoretical limits of environmentalism amplifies the constructivist tendency to produce discourses about discourses leaving no possibility for engaging with the material aspects of the issue.

Both approaches focus on the negotiation between green diplomacy, environmental organizations, and business and advocate for public regulatory power of national and international institutions (Hayward 1994). Consequently, they do not deal with the transformation of practical, daily environmental commitment, and while offering an insightful analysis, they do not challenge the current worldview.

However, the flourishing debate on post-environmentalism signals the need for a turning point in both the understanding and practice of environmentalism. Building upon this urgency, this book proposes that we are actually already living in an age of postenvironmentalism (without hyphen) rather that an age of post-environmentalism. The difference consists in the very fact that while the second is characterized by the critical consideration of what mainstream environmentalism was, the first is an affirmative approach able to reconsider—and, hopefully, to dissolve—the distinction between environmental and social issues. By starting from material rather than discursive practices, it is possible to step out the dualist frame that has long forced environmental thinkers in a never-ending debate in which nature-related concerns were opposed to society-related concerns. The perspective offered by material semiotics (also referred to as actor-network theory (ANT)) can be of help for appreciating real-life postenvironmentalist engagement and for designing a "third way" for future environmental thinking and agency (Moll and Law 2002).

A material-semiotic approach starts from the recognition that everything is generated from and located in a network of relations linking together humans, nonhumans (e.g., animals, plants, bacteria, geological formations, etc.) and more-than-humans (e.g., technologies, laws, mechanical devices, procedures, etc.); as John Law clearly states: "nothing has reality or form outside the enactment of … relations" (Law 2007, 1). Both humans and nonhumans are regarded as capable of motivated actions. Therefore, agency is not merely a human prerogative, but it is rather a property distributed through the network (Whatmore 2003). Again, it is the very existence of this network that makes agency possible because, as geographer Jonathan Murdoch explains, it is only through the network that "actors make any impact upon the world; no actor can make any kind of effective intervention without the support of others; action is association" (Murdoch 2006, 74). When heterogeneous actors assemble around a matter of concern that is relevant in the public space and take concrete action, their agency produces politically relevant consequences (Barry 2001; Bennet 2010). This does not mean that nonhuman actors can vote or sit in parliament or perform any other activity conventionally considered distinctively human; rather it more radically suggests that those activities are not the only ways to deal with political issues.

6.3   POSTENVIRONMENTALIST AGENCY IN LIVING SPACES

From a theoretical perspective, *post-environmentalism* differs from *postenvironmentalism* because the first adopts a realist or a constructivist perspective, while the latter builds upon the postmodern material-semiotic theory.

Post-environmentalism in the realist version can be regarded as an evolution of mainstream environmental thinking, which refers to universal values and objective knowledge, and provides few possibilities for local actors to affect the global networks of knowledge and power production and circulation; in the constructivist version, it challenges collective mentality by proposing new environmental values to reverse political, economic, and social order through discursive practices.

Postenvironmentalism, in turn, has (apparently) more modest aspirations. It does not propose a substantive epistemological or political theory, nor it provides a list of "must-do things" or values to be adopted; it does not define any ethical norms to create consensus on normative principles but instead reveals the reasons for environmental disputes to emerge—i.e., how, why, and where environmental issues become a matter of concern for heterogeneous collectives. It does not derive legitimization from scientific authority (whose suggestions *have to* be followed) or cultural norms (whose strength lies in the appeal of ideologies) but focuses on the collectives' micro-politics emerging from the trivial, material practices of everyday life that are able to illuminate—and eventually influence—the macro-politics generated by global power relationships (Braun and Whatmore 2010). In so doing, postenvironmentalism advances an alternative, nonrepresentational, nondiscursive, materiality-oriented view of environmental thinking and practices performed in local spaces. While rejecting the objective of reaching universal consensus on one-size-fits-all solutions, it also makes evident that when we talk of environmental issues we actually cannot talk of anything else than of the *politics* of the environment, and thus every attempt at de-politicizing environmental issues will ultimately lead us to betraying their very nature. At the same time, however, the formation of environmental issues turns out to be a much broader and protracted affair than simply introducing them into the classic political sphere; it requires revolving the political sphere itself through continuous negotiations that constitute and reconstitute matter and space (Featherstone 2008).

The appreciation of the social consequences of the work of (materially and semantically) making and unmaking the world we inhabit together with nonhuman fellows publicly exposes the political character of environmental thinking. Thus, postenvironmentalism offers a standpoint for revealing the entanglement between knowledge generation and power production processes, and the opportunity of changing things by doing things.

It can be affirmed that postenvironmentalism is first of all a new framework for understanding contemporary world that allows to engage socio-natural-technological network in environmental struggles by adopting multiple means and processes. The gathering practice decenters social agency from the discursive domains and bring practices back to the core of public commitment. There are things that bring dispersed geographies closer together, there are techniques that embody political commitment, and there are places where political decisions are actively taken through the agency of networks, although still unnoticed (see Fig. 6.1). Examples have been provided in Chap. 5 to illustrate how postenvironmentalism can offer a new perspective and new possibilities of engagement in the panorama of environmental thinking; the material-semiotic perspective makes it clear how these practices blur the boundaries between real and virtual, natural

Fig. 6.1 A child's representational drawing of the ANT mediation and translation processes in ANT. Birds, clouds, stars, and airplanes are in the sky; flowers and trees root in the earth; and worms move beneath the earth. A magician-scientist stands with tools in her hands to listen to nonhuman "speeches" and speak with the nonhuman world on her turn

Source: The author

and artificial, and how network thinking can help environmentalism in advancing innovative perspectives and possibilities for understanding and changing the world.

Postenvironmentalism moves from the purpose of transforming our understanding and relationship with the environment by assembling purpose-oriented networks around a practical matter of concern (rather than around ideological proclaims). Environmental issues need to be thought as the effects of continuous, multiple, practical, and material negotiations between different networks of actors. These negotiations obviously may well involve words but are not discursive in kind, as different agents with different forms of expression may take part in the negotiation: the declarations of international meetings, strikes, geological resilience, machines working or not working under specific weather conditions, administrative resistance to rules changing, animal species' proliferating or becoming extinct, alterations in the chemical composition of water, infections arising in cultivated crops, etc.

In Chap. 5, we discussed collaborative urban initiatives that transform space through the deployment of information and communication tools, as well as urban gardening projects. These exemplify how new social actors (i.e., the heterogeneous networks performing postenvironmentalist practices) are entering the global and postmodern public space, and how they are increasingly attracting the interest of traditional environmental actors (including environmental associations and international institutions).

The first set of initiatives refers to the transformation of urban material infrastructures using the capability of the social web to bring together collective knowledge. The material entwining and enfolding electronic and high-tech media is turning humans, nonhumans, and technologies into co-agents in environmental disputes. In particular, the crowdsourcing cases discussed mobilize technological devices and processes for linking together heterogeneous social actors in order to negotiate a broad range of environmental issues; these processes involve communities in the production of environmental knowledge and affect the related urban governance decision-making processes. The urban gardening practices refer, on the other hand, to the alliance of humans and nonhumans (i.e., plants, insects, fungi, soil, weather, etc.) in the city is greening urban public spaces and advancing virtuous sociopolitical processes through a broad array of non-formalized practices that provide residents with a sense of togetherness and purposiveness.

A material-semiotic postenvironmentalism, thus, interrogates, analyzes, and supports everyday environmentalist practices in their process of becoming, in order to reasonably and practicably achieve the goal of living together in a more sustainable way, by sharing our life space with a manifold of (sometimes) unexpected heterogeneous fellows. This reveals that environmental thinking never died; it just transformed into a vibrant, committing, and challenging call for inventive practices that unveil new possibilities for dwelling in common our planet.

REFERENCES

Adger, W., Benjaminsen, T., Brown, K., and Svarstad, H. 2001. "Advancing a Political Ecology of Global Environmental Discourses." *Development and Change* 32: 681–715.

Agyeman, J. 2005. *Sustainable Communities and the Challenge of Environmental Justice*. New York: New York University Press.

Alcàntara, A.M. *The End of Exclusive Environmentalism*. Ensia. http://ensia.com/voices/the-end-of-exclusive-environmentalism/ July 2, 2013.

Alier, J. 2002. *The Environmentalism of the Poor: A Study of Ecological Conflicts and Valuation*. Northampton (MA): Edward Elgar Publishing.

Arias Maldonado, M. 2012. *Real Green: Sustainability After the End of Nature*. Farnham: Ashgate Publishing.

Arias Maldonado, M. 2015. "Environment & Society. Socionatural Relations." In *The Anthropocene*. New York: Springer.

Baker, I., 1991. "Radical Conservatism," *Future*, March: 204–206.

Bailey, R. 2012. "Postenvironmentalism and Technological Abundance - A Review of Love Your Monsters, a Collection of Essays on a New Kind of Environmentalism." Reason Magazine. January 4. http://reason.com/archives/2012/01/04/postenvironmentalism-and-technological-a.

Bailey, S. 2012. "Postenvironmentalism and Technological "Abundance Reason." Accessed January 4, 2012. http://reason.com/archives/2012/01/04/postenvironmentalism.

Barbier, E. 2010. *A Global Green New Deal*. Cambridge: Cambridge University Press.

Barry, A. 2001. *Political Machines. Governing a Technological Society*. London: The Athlone Press.

Bartollomei, S. 1995. *Etica e natura*. Bari: Laterza.

Bate, R. 1995. "Post-environmentalism." *Economic Affairs* Autumn.

© The Author(s) 2016
C. Certomà, *Postenvironmentalism*,
DOI 10.1057/978-1-137-50790-7

125

Bec, U. 1995. *Ecological Politics in an Age of Risk.* Cambridge: Polity Press. Originally published in *Gegengifte: Die organisierte Unverantwortlichkeit,* 1988.

Beck, U. 1995. *Ecological Politics in an Age of Risk.* Cambridge: Polity press.

Beckie, M., and Bogdan, E. 2010. "Planting Roots: Urban Agriculture for Senior Immigrants." *Journal of Agriculture. Food Systems and Community Development.* 1/2: 77–89.

Been, V., and Voicu I. 2006. "The Effect of Community Gardens on Neighboring Property Values." New York University Law and Economics Working Papers, 46.

Beevers, M.D., and Petersen, B.C. 2009. "Review of T. Nordhaus and M. Shellenberger's Break through." *Society & Natural Resources: An International Journal* 22:783–785.

Bennet, J. 2001. *The Enchantment of Modern Life. Attachments, Crossing and Ethics.* Princeton: Princeton University Press.

Bennet, J. 2010. *Vibrant Matter. The Political Ecology of Things.* Durham and London: Duke University Press.

Berg, P. 2001. "The Post-Environmentalist Directions of Bioregionalism." Lecture by Peter Berg. Missoula: University of Montana. Accessed April 10, 2001. http://www.planetdrum.org/Post-Enviro.htm.

Berke, P.R., and Conroy, M.M. 2000. "Are We Planning for Sustainable Development?." *Journal of the American Planning Association* 66:21–33.

Biermann, F. 2006. "Global Governance and the Environment." In *International Environmental Politics* edited by M. Betsill, K. Hochstetler and D. Stevis, 237–261. Basingstoke (UK): Palgrave Macmillan.

Bingham, N. 2006. "Bees, Butterflies, and Bacteria: Biotechnology and the Politics of Nonhuman Friendship". *Environment and Planning A* 38: 483–498.

Bingham, N., and Hinchliffe, S. 2008. "Reconstituting Natures: Articulating Other Modes of Living Together". *Geoforum* 39: 83–87.

Bled, A. 2010. "Technological Choices in International Environmental Negotiations: An Actor." *Network Analysis Business Society* 49: 570–590.

Blühdorn 1997. "Ecological Thought and Critical Theory." In *Green Thought in German Culture* edited by C. Riordan. Cardiff: University of Wales Press.

Blühdorn, I. 2000. *Post-Ecologist Politics. Social Theory and the Abdication of the Ecologist Paradigm.* London & New York: Routledge.

Blühdorn, I. 2006. "Self-Experience in the Theme Park of Radical Action? Social Movements and Political Articulation in the Late-modern Condition." *European Journal of Social Theory* 9: 23–42.

Blühdorn, I. 2007a. "Sustaining the Unsustainable: Symbolic Politics and the Politics of Simulation." *Environmental Politics* 16/2: 251–275.

Blühdorn, I. 2007b. "The Politics of Unsustainability: Eco-Politics in the Post-Environmental Era." In *Environmental Politics (Special Issue) Vol. 16*: 185 edited by I. Blühdorn and L. Wels. London: Routledge.

Blühdorn, I. 2011. "The Politics of Unsustainability: COP15, Post-ecologism, and the Ecological Paradox." *Organization and Environment* 24: 34–53.

Blühdorn, I., and Welsh, I. 2007. "Eco-politics Beyond the Paradigm of Sustainability: A Conceptual Framework and Research Agenda." *Environmental Politics* 16: 185–205.

Blühdorn, I., and Welsh, I. 2008. *The Politics of Unsustainability: Eco-Politics in the Post-Environmental Era.* London & New York: Routledge.

Bookchin, M. 1982. *The Ecology of Freedom: The Emergence and Dissolution of Hierarchy.* Palo Alto: Cheshire Books.

Boyle, A., Redgwell, C., and Birnie, P. 2009. *International Law and the Environment.* 3rd edn. New York: Oxford University Press.

Brabham, D.C. 2009. "Crowdsourcing the Public Participation Process for Planning Projects." *Planning Theory* 8: 242–262.

Brabham, D.C. 2012. "Motivations for Participation in a Crowdsourcing Application to Improve Public Engagement." *Transit. Planning.* 40: 307–328.

Bramwell, A. 1989. *Ecology in Twentieth Century: A History.* New Haven & London: Yale University Press.

Braun, B., and Castree, N. 1998 *Remaking Reality: Nature at the Millennium.* London: Routledge.

Braun, B., and Wainwright, J. 2001. "Nature, Poststructuralism, and Politics." In *Social Nature, Blackwell* edited by N. Castree and B. Braun. Oxford: Blackwell.

Braun, B., and Whatmore, S.J. 2010. *Political Matter: Technoscience, Democracy and Public Life.* Minneapolis: University of Minnesota Press.

Brick, P., and Cawley, R.M. 2008. "Producing Political Climate Change: The Hidden Life of US Environmentalism." *Environmental Politics* 17: 200–218.

Broder, J. 2011. "An Energy Plan Derailed by Events Is Being Retooled," *The New York Times,* March 30.

Bromberg, A., Morrow, G.D., and Pfeiffer, D. 2007. *Critical Planning.* Summer.

Brosius, J.P. "Endangered Forest, Endangered People: Environmentalist Representation of Indigenous Knowledge." *Human Ecology* n 25.

Brosius, P. 1997. "Prior Transcripts, Divergent Paths: Resistance and Acquiescence to Logging in Sarawak, East Malaysia." *Comparative Studies in Society and History* 39.

Bryant, R.L., and Baile, S. 1997. *Third World Political Ecology.* London: Routledge.

Buck, C. 2012. "Post-environmentalism: An Internal Critique." *Environmental Politics* 22/6. DOI:10.1080/09644016.2012.712793.

Buell, F. 2003. *From Apocalypse to Way of Life: Environmental Crisis in the American Century.* London: Routledge.

Carson, R. 1962. *Silent Spring.* Cambridge (MA): Riverside Press.

Castells, M. 1997. *The Power of Identity, The Information Age: Economy, Society and Culture.* Cambridge (MA), Oxford (UK): Blackwell.

Castells, M. 1998. *The Power of Identity: The Information Age–Economy, Society and Culture.* Oxford: Blackwell Publishing.

Castree, N. 2006. "The Future of Environmentalism." *Soundings* n. 34, Nov.

Castree N. 2001. "Socialising Nature: Theory, Practice, and Politics." In Social Nature, Blackwell edited by N. Castree and B. Braun. Oxford: Blackwell.

Castree N., and MacMillan T. 2001. "Dissolving Dualisms: Actor-Network and the Reimagination of Nature." In *Social Nature* edited by N. Castree and B. Braun. Oxford: Blackwell Publishing.

Centro di Documentazione sui Conflitti Ambientali (CDCA). 2011. "Estrazioni petrolifere nel parco nazionale dello Yasuni." Accessed July 11, 2011. http://www.cdca.it/spip.php?article606.

Certomà, C. 2006. Ecology, Environmentalism and System Theory. *Kybernetes* 35(6): 915–921.

Certomà, C. 2009. "Environmental Politics and Place Authenticity Protection." *Environmental Values* 18: 313–341.

Certomà, C. 2011. "Critical Urban Gardening as a Post-Environmentalist Practice." *Local Environment* 16(10): 977–987.

Certomà C., and Greyl, L. 2012. "Non-Extractive Policies as a Main Road to Environmental Justice? The Case of the Yasunì Park in Ecuador." In *New Political Spaces in Latin American Natural Resource Governance* edited by H. Haarstad. New York (NY): Palgrave Macmillan.

Certomà, C., Corsini, F. and Rizzi, F. 2015. "Crowdsourcing Urban Sustainability. Data, People and Technologies in Participatory Governance." *Futures* (forth. 2015).

Chaloupka, W. 2008. "The Environmentalist: 'What Is to Be Done?'." *Environmental Politics* 17:237–253.

Christoff, P. 1996. "Ecological Modernisation, Ecological Modernities." *Environmental Politics* 5/3: 209–231.

Christoph, T. 2010. *Genetic Engineering Enforces Corporate Control of Agriculture.* Greenpeace.http://www.greenpeace.org/international/en/publications/reports/GE-enforces-corporate-control-of-agriculture/.

Clark, D., and Chadwick, M. 2011. *The Rough Guide to Community Energy.* London: Rough Guides.

Collectif Argos. 2010. *Climate Refugees.* Cambridge (MA): MIT Press.

Collina, T., and Poff, E. 2009. "The Green New Deal: Energizing the U.S." *Economy Fokus America 4.* http://library.fes.de/pdf-files/bueros/usa/06873.pdf.

Colombo, L. 2004a. "Consiglio dei diritti genetici." In *Il grano transgenico:un evento inatteso.* http://www.fondazionedirittigenetici.org/fondazione/new/displaystudio.php?id=194.

Colombo, L. 2004b. *Grano o grane La sfida OGM in Italia.* Lecce: B Manni S. Cesario.

Commoner, B. 1971. *The Closing Circle: Nature, Man, and Technology.* New York: Knopf.

Cook, I. 2004. "Follow the Thing: Papaya." *Antipode* 36: 624–664.

Cultural survival. 2015. "The Thinking People: The U'wa Battle Oxy." http://www.culturalsurvival.org/publications/cultural-survival-quarterly/colombia/thinking-people-uwa-battle-oxy.

Daly, H., and Farley, J. 2004. *Ecological Economics. Principles and Applications.* Washington (DC): Island press.

Darier, E. 1999. *Discourses of the Environment.* Oxford: Blackwell.

David, G., and Victor. 2001. *The Collapse of the Kyoto Protocol and the Struggle to Slow Global Warming.* Princeton: Princeton University Press.

Davidson, D.J. 2009. "Review of T. Nordhaus and M. Shellenberger's Break Through." *Sustainability: Science, Practice, and Policy* 5. http://sspp.pro quest.com/archives/vol5iss1/book.nordhaus.html.

De Vosab, M.G, Janssen, P.H.M., Kok, M.T.J., Frantzi, S., Dellas, E., Pattberg, P., Petersen, A.C., and Biermann, F. 2013. "Formalizing Knowledge on International Environmental Regimes: A First Step Towards Integrating Political Science in Integrated Assessments Of Global Environmental Change." *Environmental Modelling & Software* 44: 101–112.

Deleuze, G., and F. Guattari. 2002. *A Thousand Plateaus: Capitalism and Schizophrenia.* Minneapolis: University of Minnesota Press.

Descola, P. 1996. "Constructing Nature. Symbolic Ecology and Social Practice." In *Nature and Society Anthropological Perspectives* edited by P. Descola and G. Palsson. London: Routledge.

Despret, V. 2005. "Sheep Do Have Opinions." In *Making Things Public-Atmosphere of Democracy* edited by B. Latour and P. Weibel. Cambridge (MA): MIT Press.

Devinney, T. June 22nd, 2012, 5.24am BST. "Why the Global Environmental Movement is Failing." In *The Conversation.* https://theconversation.com/why-the-global-environmental-movement-is-failing-7819.

Dobson, A. 1990. *The Green Reader.* London: Andre Deutsch.

Dobson, A. 1998. *Justice and the Environment: Conceptions of Environmental Sustainability and Theories of Distributive Justice.* Oxford: Oxford University Press.

Dobson, A. 2003a. "Ecologism and Environmentalism." In *Contemporary Political Thought: A Reader and Guide Contemporary Political Thought: A Reader and Guide* edited by A. Finlayson. Edinburgh: Edinburgh University Press.

Dobson, A. 2003b. *Citizenship and the Environment.* New York: Oxford University Press.

Dobson, A. 2010. "Democracy and Nature: Speaking and Listening." *Political Studies* 58: 752–768.

Donella, H., Meadows Dennis, L., Randers, J., and Behrens II, W.W. 1972. *Limits to Growth.* New York: New American Library.

Donoso, A. 2009. "Game DEUDA ECOLÓGICA Impactos de la deuda externa en las comunidades y la naturaleza" *Accion Ecologica Quito – Ecuador*, September.

http://www.accionecologica.org/deuda-ecologica/documentos/1177-nueva-publicacion-qdeuda-ecologica-impactos-de-la-deuda-externa-en-las-comuni dades-y-la-naturalezaq.

Dresner, S. 2002. *The Principle of Sustainability.* London: Earthscan publication Ltd.

Dryzek, J. 1987. *Rational Ecology: Environment and Political Economy.* New York: Basil Blackwell.

Dube, S. 2002. "Introduction: Enchantments of Modernity." *The South Atlantic Quarterly* 101: 729–755.

Duncan, C., and Chadwick, M. *The Rough Guide to Community Energy.* London: Penguin.

Eder, K. 1996a. "The Institutionalisation of Environmentalist." In *Risk, Environment and Modernity* edited by S. Lash, B. Szerszynsky and B. Wynne. London: Sage.

Eder, K. 1996b. *The Social Construction of Nature.* London: Sage.

Eder, K. 1996c. *The Social Construction of Nature.* Nottingham on Trent: Sage Publications. Originally published in Die Vergesellshaftung der Natur: Studien zur sozialen Evolution der praktishen Vernunft. 1988.

Ehrlich, P. 1968. *The Population Bomb.* New York: Sierra Club/Ballantine Books.

EJOLT. 2015. "Environmental Justice Organisation, Liability and trade." EU FP/7, http://www.ejolt.org/.

El-Hinnawi, E. 1985. "Environmental Refugees." UNEP, Nairobi.

Ellis, E. 2011. "Planet of No Return: Human Resilience on an Artificial Earth." In *Love Your Monsters: Postenvironmentalism and the Anthropocene* edited by M. Shellenberg and T. Nordhaus. Washington D.C.: The Breacktrought Institute.

Emmett, R. 2011. "Community Gardens, Ghetto Pastoral, and Environmental Justice." *Interdisciplinary Studies in Literature and Environment* 18/1: 67–86.

Escobar, A. 1999. "Gender, Place and Networks. A Political Ecology of Cyberculture." In *Women@Internet. Creating New Cultures in Cyberspace* edited by W. Harcourt. London: Zed Books.

Faburel, G. 2010. "The Environment as a Factor of Spatial Injustice: A New Challenge for the Sustainable Development of European Regions?." In *Sustainable Development–Policy and Urban Development* edited by C. Ghenai. InTech. Accessed January 3, 2013. http://www.intechopen.com/books/.

Fairhead J., and Leach, M. 1996. *Misreading African Landscape Society and Ecology in Forest-Savanna Mosaic.* Cambridge: Cambridge University Press.

Fall, J.J. 2005. *Drawing The Line: Nature, Hybridity And Politics In Transboundary Spaces.* London: Ashgate.

Fall, J.J. 2014. "Biosecurity and Ecology: Beyond the Nativist Debate." In *Biosecurity: The Socio-Politics of Invasive Species and Infectious Diseases* edited by K. Barker, A. Dobson and S. Taylor. Abingdon: Earthscan/ Routledge.

Featherstone, D. 2002. *Spatiality, Political Identity and the Environmentalism of the Poor.* Unpublished PhD thesis. The Open University.

Featherstone, D.J. 2008. *Resistance, Space and Political Identities: The Making of Counter-Global Networks.* Chichester: Wiley-Blackwell.

Ferris J, Norman, C., and Sempik, J. 2001. "People, Land and Sustainability: Community Gardens and the Social Dimension of Sustainable Development." *Social Policy Administration* 35/5: 559–568.

Fischer, C., Parry, I., Aguilar, F., and Jawahar, P. 2005. "Conducted for the Foreign Investment Advisory Service of the World Bank Group Corporate Codes of Conduct: Is Common Environmental Content Feasible?" *Resources for the Future.* Discussion Paper, March 05–09.

Flachs A. 2010. "Food For Thought: The Social Impact of Community Gardens in the Greater Cleveland Area." *Electronic Green Journal* 1/30: 1–9.

Forsyth, T. 2008. "Political Ecology and the Epistemology of Social Justice." *Geoforum* 39: 756–764.

Foucault, M. 1977. *Discipline and Punish. The Birth of the Prison,* Penguin, London. Originally published in. Naissance de la prison. 1975.

Foucault M. 2003. "Society Must Be Defended." *Lectures at the College de France 1975–76.* New York: Picador.

Fraser, N. 1997. *Justice Interruptus: Critical Reflections on the "Postsocialist" Condition.* New York: Routledge.

Gabrys, J. 2011. *Digital Rubbish: A Natural History of Electronics.* Ann Arbor: University of Michigan Press.

Geertz, C. 2000. *Available Light: Anthropological Reflections on Philosophical Topics.* Princeton: Oxford Princeton University Press.

Giddens, A. 1990. *The Consequences of Modernity.* Stanford: Stanford University Press.

Gillis J. 2003. "Farmers Divided Over Introduction of GE Wheat." *Washington Post,* April 22.

Gleason, H. 1939. "The Individualistic Concept of the Plant Association." *American Midland Naturalist* n 21.

Gollier, C., Jullien, B., and Treich, N. 2000. "Scientific Progress and Irreversibility: An Economic Interpretation of the 'Precautionary Principle'." *Journal of Public Economics* 75 (2): 229–253.

Gore, A. 2006. "Averting the Climate Crisis." TED. Filmed February.

Gore, A. 2008. *The Assault of Reason.* New York: Penguin Group.

Gore, A. 2009. *Our Choice: A Plan to Solve the Climate Crisis.* New York: Penguin Group.

Government of Ecuador. 2010. "Ecuador Yasuni ITT Trust Fund: Terms of Reference." Accessed March 1, 2012. http://mdtf.undp.org/document/download/4492.

Government of Ecuador. 2011. "El Parque Yasuní: el más biodiverso del mundo." *Yasuni ITT. Una iniciativa por la vida.* Accessed July 16, 2011. http://yasuni-itt.gob.ec/%C2%BFpor-que-ecuador-propone-la-iniciativa-yasuni-itt/conservar-la-biodiversidad/la-biodiversidad-del-parque-nacional-yasuni/el-par que-yasuni-el-mas-biodiverso-del-mundo/.

Green New Deal. 2014. "What Is the Green New Deal?." http://greennewdeal.eu/what-is-the-green-new-deal.html.

Greimas, A.J. 1986. *Sémantique structurale.* Paris: Presse universitaires de France (or.ed. 1966).

Grenier, M. 2002. "Agronomic Assessment of Roundup Ready Wheat." CWB discussion paper.www.cwb.ca.

Grove, R.H. 1995. *Green Imperialism: Colonial Expansion, Tropical Island Edens and the Origins of Environmentalism.* Cambridge: Cambridge University Press.

Habermas, J. 1984. "Reason and the Rationalization of Society." *Volume 1 of The Theory of Communicative Action.* English translation by Thomas McCarthy. Boston: Beacon Press.

Haq, G., and Alistair, P. 2011. *Environmentalism Since 1945.* London: Routledge.

Haraway, D. 1991. "A Cyborg Manifesto Science, Technology, and Socialist-Feminism in the Late Twentieth Century." In *Simians, Cyborgs and Women: The Reinvention of Nature,* 149–181. New York: Routledge.

Haraway, D. 1992. "The Promises of Monsters: A Regenerative Politics for Inappropriate/d Others." In *Cultural Studies* edited by L. Grossberg, C. Nelson and P. A. Treichler. New York: Routledge.

Haraway, D. 1997. *Modest Witness@Second Millennium. Female Man Meets_OncoMouse™–Feminist and Technoscience.* New York: Routledge.

Harvey, D. 1973. *Social Justice and the City.* Baltimore: Johns Hopkins University Press.

Harvey, D. 1996. *Justice, Nature and the Geography of Difference.* Oxford: Basil Blackwell.

Haughton, G. 1999. "Environmental Justice and the Sustainable City." *Journal of Planning Education and Research* 18: 233–243.

Hay, P. 2002. *Main Currents in Western Environmental Though.* Bloomington (IN): Indiana University Press.

Hayward, T. 1994. *Environmental Thought.* Cambridge: Polity Press.

Hayward, T. 2003. "Ecologism and Environmentalism." In *Contemporary Political Thought: A Reader and Guide Contemporary Political Thought: A Reader and Guide* edited by A. Finlayson. Edinburgh: Edinburgh University Press.

Heal, G. 2000. *Nature and the Marketplace: Capturing the Value of Ecosystem Services.* Washington: Island Press.

Henson, R. 2011. *The Rough Guide to Climate Change.* London: Penguin.

Hinchcliffe, S. 2003. *Inhabiting — Landscapes and Natures.* In *Handbook of Cultural Geography* edited by K. Anderson, M. Domosh, S. Pile and N. Thrift. London: Sage.

Hinchliffe, S. 2007. *Geography of Nature: Societies, Environments, Ecologies.* London: Sage.

Hinchliffe, S, and Whatmore, S. 2006. "Living Cities: Toward a Politics of Conviviality." *Science as Culture* 15: 123–138.

Homer-Dixon, T.F. 2001. *Environment, Scarcity, and Violence.* Princeton (NJ): Princeton University Press.

Hopkins, R. 2013. *The Power of Just Doing Stuff: How Local Action Can Change the World.* Cambridge: Green Books.

Hou J., Johnson, J., and Lawson, L. 2009. "Greening Cities Growing Communities: Learning from Seattle's." *Urban Community Gardens.* Washington & London: University of Washington Press.

Huggan, G., and Tiffin, H. 2010. *Postcolonial Ecocriticism. Literature, Animals, Environment.* New York: Routledge.

Horgan, J. 2011. "Killing Environmentalism to Save It: Two Greens Call for 'Postenvironmentalism'." *Scientific American.* December 26. http://blogs.scientificamerican.com/cross-check/killing-environmentalism-to-save-it-two-greens-call-for-postenvironmentalism.

IPCC. 2007. "Summary for Policymakers." A report of Working Group I of the Intergovernmental Panel on Climate Change.

Istanbul Declaration and The Habitat Agenda. 1996. "Istanbul Declaration on Human Settlements."

Kates, R., Parris, T., and Leisorowitz, A. 2005. "What Is Sustainable Development?." *Environment* 47: 8–21.

Katz, C. 1998. "Whose Nature, Whose Culture? Private Productions of Space and the 'Preservation' of Nature'." In *Remaking Reality: Nature at the Millennium* edited by B. Brawn and N. Castree. London: Routledge.

Ksentini, Z. 1994. "Human Rights and the Environment." *UNHRC.* E/CN.4/Sub.2/1994/9.

Kurtz, H. 2001. "Differentiating Multiple Meanings of Garden and Community." *Urban Geography* 22: 656–670.

Kysar, D.A. 2008. "The Consultants' Republic." *Harvard Law Review* 121: 2041–2084.

Laclau, E., and Mouffe, C. 1985. *Hegemony and Socialist Strategy: Towards a Radical Democratic Politics.* London: Verso.

Larrea, C., and Warnars, L. 2009. "Ecuador's Yasuni-ITT Initiative: Avoiding Emissions by Keeping Petroleum Underground." *Energy for Sustainable Development* 13: 219–223.

Latour, B. 1991. "Society Is Technology Made Durable." In *A Sociology of Monsters: Essays on Power, Technology and Domination* edited by J. Law, 103–132. New York/London: Routledge.

Latour, B. 1993a. *The Pasteurization of France.* Cambridge (MA): Harvard University Press.

Latour, B. 1993b. *We Have Never Been Modern*. Cambridge (MA): Harvard University Press.

Latour, B. 1996. *Aramis or the Love of Technology*. Cambridge (MA): Harvard University Press.

Latour, B. 1999. "On Recalling ANT." In *Actor-Network Theory and After* edited by Law and Hassard, 15–26. Oxford: Blackwell Publishers.

Latour, B. 2004. *Politics of Nature: How to Bring the Sciences into Democracy*. Cambridge (MA): Harvard University Press.

Latour, B. 2005. *Reassembling the Social*. Oxford: Oxford University Press.

Latour, B. 2008. "'It's Development, Stupid !' or: How to Modernize Modernization." In *Post-environmentalism* edited by J. Proctor. Cambridge (MA): MIT Press.

Latour, B. 2011. "Love Your Monsters: Why We Must Care For Our Technologies As We Do Our Children." In *Love Your Monsters: Postenvironmentalism and the Anthropocene* edited by M. Shellenberg and T. Nordhaus. Washington D.C.: The Breacktrought Institute.

Latour, B., and Weibel, P. 2005. *Making Things Public-Atmosphere of Democracy*. Cambridge (MA): MIT Press.

Law, J. 1991. "Strategies of power. Power, discretion and strategy." In *A Sociology of Monsters: Essays on Power, Technology and Domination* edited by J. Law, 165–191. New York/London: Routledge.

Law, J. 1995. "Introduction: Monsters, Machines and Sociotechnical Relations." In *The Sociology of Science* edited by H. Nowotny and Taschwer. London: Edward Elgar, (reprint).

Law, J. 2004. "Enacting Naturecultures: A Note from STS." Published by the Centre for Science Studies. Lancaster: Lancaster University. http://www.comp.lancs.ac.uk/sociology/papers/law-enacting-naturecultures.pdf.

Law, J. 2007. *Actor Network Theory and Material Semiotics*. Lancaster. Lancaster University Centre for Science Studies. http://www.lancaster.ac.uk/fass/centres/css/ant/ant.htm.

Law. 2008. "Actor-Network Theory and Material Semiotics." In *The New Blackwell Companion to Social Theory*. 3rd edn. edited by B.S. Turner, 141–158. Oxford: Blackwell.

Law, J., and Hetherington, K. 2003. "Materialities, Spatialities, Globalities." in *The Spaces of Postmodernism: Readings in Human Geography* edited by M. Dear and M. Flusty, 390–401. Oxford: Blackwell.

Lefebvre, H. 1991. *The Production of Space*. Oxford: Basil Blackwell. Originally published in 1974.

Leonard L., and Kedzior, S. 2014. *Occupy the Earth: Global Environmental Movements Emerald Group*. Bingley: Emerald Group Publishing Limited.

Leopold, A. 1968. *A Sand Country Almanac*. Oxford: Oxford University Press.

Lomborg, B. 2001. *The Skeptical Environmentalist: Measuring the Real State of the World*. Cambridge: Cambridge University Press.

Luke, T.W. 1998. "Environmentality as Green Governmentality." In *Discourses of the Environment* edited by E. Darier. Oxford: Wiley-Blackwell.

Luke, T.W. 1999. "Environmentality as Green Governmentality" In *Discourses of the Environment* edited by E. Darier. Oxford: Blackwell.

Luke, T.W. 2009. "An Apparatus of Answers? Ecologism As Ideology in the 21st Century." *New Political Science* 31: 487–498.

Lyotard, J.-F. 1984. *The Postmodern Condition*. Manchester: Manchester University Press (or. ed. La Condition postmoderne: rapport sur le savoir, 1979).

Manser, B. 1996. "Voices from the Rainforest: Testimonies of a Threatened People." http://www.survivalinternational.org/tribes/penan.

March, H., and Ribera, R. 2014. "Smart Contradictions: The Politics of Making Barcelona a Self-Sufficient City." *European Urban and Regional Studies* 1–15. ISSN.0969-7764.

Margalef, R. 1977. *Ecologìa*. Barcelona: Omega. Lovelock, J.E. 1979. *Gaia: A New Look at Life on Earth*. Oxford: Oxford University Press.

Marres, N. 2005. "Issues Spark a Public Into Being. A key but Often Forgotten Point of the Lippmann-Dewwey Debate." In *Making Things Public-Atmosphere of Democracy* edited by B. Latour and P. Weibel. Cambridge (MA): MIT.

Marres, N. 2012. *Material Participation. Technology, the Environment and Everyday Publics*. New York: Palgrave Macmillan.

Marres N., and Rogers, R. 2005. "Recipe for Tracing the Fate of Issues and Their Publics on the Web." In *Making things Public-Atmosphere of Democracy* edited by B. Latour and P. Weibel. Cambridge (MA): MIT.

Martinez Alier, J. 2003. *The Environmentalism of the Poor: A Study of Ecological Conflicts and Valuation*. Cheltenham (UK), Northampton (MA): Edward Elgar Publishing.

Mary Louise Pratt. 1992. *Imperial Eyes: Travel Writing and Transculturation*. London: Routledge.

Maskit. 1998. "Something Wild? Deleuze and Guattari and the Impossibility of Wilderness." In *Philosophies of Place* edited by A. Light and J.M. Smith. Lanham (MD): Rowman & Littlefield Publishers.

Massey, D. 2006. Landscape as a Provocation. Reflection on Moving Mountains. *Journal of Material Culture* 11: 33–48.

Mauz, I, and Gravelle, J. 2005."Wolves in the Valley. On making a Controversy Public." In *Making Things public-Atmosphere of Democracy* edited by B. Latour and P. Weibel. Cambridge (MA): MIT Press.

Max, W. 2005. *The Protestant Ethic and the Spirit of Capitalism*. London: Routledge.

Mayr, E. 1997. *This is Biology -The Science of the Living World*. Cambridge (MA): The Belknappress of Harvard University Press.

McClintock, N. 2008. "From Industrial Garden to Food Desert: Unearthing the Root Structure of Urban Agriculture in Oakland, California." Institute for Study of Societal Issues Working Papers. http://escholarship.org/uc/item/1wh3v1sj.

McCormick, J. 1991. *British Politics* and the *Environment*. London: Earthscan.

McCrory Martin, A., and Langvardt Kyle, T. 2012. "Cutting Out the Middle-Man: The Case for Direct Business Involvement in Environmental Justice." *Business Horizons* 55: 357–362.

Mebratu, D. 1998. "Sustainability and Sustainable Development: Historical and Conceptual Review." *Environmental Impact Assessment Review.*18: 493–520.

Melvin, J. 2009. "Obama's Green Energy Plans Build Hopes, Scepticism." *Reuters*, January 9.

Mennis, J. 2002. "Using Geographic Information Systems to Create and Analyse Statistical Surfaces of Population and Risk for Environmental Justice Analysis." *Social Science Quarterly* 83: 281–297.

Meyer, J.M. 2005. "Does Environmentalism Have a Future?." 69–75. *Dissent* (Spring).

Milbourne, P. 2012. "Everyday (in)justices and Ordinary Environmentalisms: Community Gardening in Disadvantaged Urban Neighbourhoods." *Local Environment* 17(9): 943–957.

Miller, J. 2005. "Biodiversity Conservation and the Extinction of Experience." *Trends in Ecology & Evolution* 20(8): 430–434.

Milton, K. 1996. *Environmentalism and Cultural Theory*. London: Routledge.

Mol, A. 2002. "The Body Multiple." In *Ontology in Medical Practice*. Durham: Duke University Press.

Mol, A., and Law, J. 2002. *Complexities*. Durhamf: Duke University Press.

Monsanto. 2003. *Bringing New Technologies to Wheat – Information on the Development of Roundup Ready Wheat*. http://www.monsanto.com.

Monsanto. 2004. "Monsanto to Realign Research Portfolio, Development of Roundup Ready Wheat Deferred." *Monsanto*, May 10. http://news.monsanto.com/press-release/monsanto-realign-research-portfolio-development-roundup-ready-wheat-deferred.

Mouffe, C. 1998. "The Radical Centre: A Politics Without Adversary." *Soundings* 9: 11–23.

Murdoch, J. 2006. *Post-Structuralist Geography*. London: Sage.

Naess, A. 1989. *Ecology, Community, and Lifestyle*. Cambridge: Cambridge University Press.

Neri, V. 2004. "FAI modelli organizzativi e dialettica istituzionale di un'associazione postambientalista." *Equilibri*. Il Mulino.

Neumayer, E. 2010. *Weak Versus Strong Sustainability: Exploring the Limits of Two Opposing Paradigms*, Edward Elgar Publishing, Incorporated.

Niklas. 1995. *Social Systems*. Stanford, CA: Stanford University Press. Adorno, Theodor W. 1973. *Negative Dialectics*. London: Routledge.

Nobelprize.org. 2007. MLA Style: "The Nobel Peace Prize 2007." Assessed 11 Jan 2015. http://www.nobelprize.org/nobel_prizes/peace/laureates/2007/.

Nordhaus, T., and Shellenberger, M. 2007. "Second Life: A Manifesto for a New Environmentalism." *The New Republic*, October. http://thebreakthrough.org/archive/second_life_a_manifesto_for_a.

Nordhaus, T., and Shellenberger, M. 2009. "The Emerging Climate Consensus: Global Warming Policy in a Post-Environmental World." *The Breacktrought Institute*. http://thebreakthrough.org/archive/the_emerging_climate_consensus.

O'Connor, J. 1998. *Natural Causes: Essays in Ecological Marxism*. New York: Guilford Press.

OECD. 2012. *OECD 2012 Environmental Outlook to 2050*. OECD Publishing. DOI:10.1787/9789264122246-en.

Olesen, F., and Markussen, R. 2007. "How to Place Material Things: From Essentialism to Material Semiotic Analysis of Sociotechnical Practice." *Convergence* 13: 79–91.

Pagano, M.A., and Bowman, A. 2000. "Vacant Land in Cities: An Urban Resource." *The Brookings Institution Survey Series*, 8.

Parikka, J. 2011. *The Materiality of Information Technology and Electronic Waste*. Cambridge (MA): MIT Press.

Pengra, B. 2012. "One Planet, How Many People? A Review of Earth's Carrying Capacity." A Discussion Paper for the Year of RIO+20. June. http://www.unep.org/geasUNEP.

Pinkerton, T., and Hopkins, R. 2009. *Local Food. How to Make It Happen in Your Community*. Devon: Green Books.

Pope, C. 2005. "An In-depth Response to "The Death of Environmentalism."" *The Grist*, January 14. http://grist.org/article/pope-reprint/.

Rawls, J. 1971. *A Theory of Justice*. Cambridge (MA): Harvard University Press.

Rees, W.E. 1995. "Achieving Sustainability: Reform or Transformation?." *Journal of Planning Literature* 9:343–361.

Rees, W.E., and Wackernagel, M. 1994. "Ecological Footprints and Appropriated Carrying Capacity: Measuring the Natural Capital Requirements of the Human Economy." In *Investing in Natural Capital: The Ecological Economics Approach to Sustainability* edited by A.M. Jansson, M. Hammer, C. Folke and R. Costanza. Washington (DC): Island Press.

Ribera, R., Santangelo, M., and Vanolo, A. 2015. "Technology and the Cities of Tomorrow: Exploring the Smart City Imagery (and Beyond)." *AAG* Chicago conference programme http://www.aag.org/.

Robbins, P. 2004. *Political Ecology*. Oxford: Blackwell Publishers.

Roberts, J. 2010. *Environmental Policy*. London: Routledge.

Rodrigues, G., and Rabben, L. 2007. *Walking the Forest with Chico Mendes*. Austin: University of Texas Press.

Roos, B., and Hunt, A. 2010. *Postcolonial Green: Environmental Politics and World Narratives*. Charlottesville: University of Virginia Press.

Rootes, C. 2008. "Review of T. Nordhaus and M. Shellenberger's Break through." *International Affairs* 84: 1317–1319.

Rorty, R. 1980. *Philosophy and the Mirror of Nature*. Princeton (NJ): Princeton University Press.

Sachs, W. 1993. "Global Ecology and the Shadow of 'Development'." In *Global Ecology: A New Arena of Political Conflict* edited by W. Sachs. Halifax (Nova Scotia): Fernwood Books.

Sachs, W. 1997. "Sustainable Development." In *The International Handbook of Environmental Sociology* edited by M. Redclif. Cheltenham: Edward Elgar.

Sagoff, M. 2011. "The Rise and Fall of Ecological Economics." In *Love Your Monsters: Postenvironmentalism and the Anthropocene* edited by M. Shellenberg and T. Nordhaus. Washington D.C.: The Breacktrought Institute.

Said, Edward. 1977. *Orientalism*. London: Penguin.

Santolini, F. 2012. "Ambientalismo 2.0." *Huffington post*, October 30. http://www.huffingtonpost.it/francesca-santolini/ambientalismo-20_b_2039843.html.

Sarewitz, Daniel. 2011. "Liberalism's Modest Proposal, Or the Tyranny of Scientific Rationality". In *Love Your Monsters: Postenvironmentalism and the Anthropocene* edited by M. Shellenberg and T. Nordhaus. The Breacktrought institute.

Sassen, S. 1998. *Globalization and Its Discontents. Essays on the New Mobility of People and Money*. New York: New Press.

Schiller, F. 2005. "Environmental Symbolism and the Discourse on Sustainability." Paper presented at conference of Munich Institution for Social Research on Sustainability, Munich, Germany.

Schlosberg, D., and Rinfret, S. 2008. "Ecological Modernisation, American style." *Environmental Politics* 17: 254–275.

Schmelzkopf, K. 1995. "Urban Community Gardens as Contested Space." *Geographical Review* 85: 364–381.

Schubert, R. 2001. "Farmers, Foreign Markets Send Negative Signals About Roundup Ready Wheat." *Cropchoice news*. http://www.cropchoice.com/lead stry923b.html?recid=228.

Schumacher, E.F. 1973. *Small is Beautiful: A Study of Economics as if People Mattered*. London: Blond&Briggs.

Sennett, Richard. 2012. "No One Likes a City That's Too Smart." *The Guardian*, December 4.

Shellenberger, M., and T. Nordhaus. 2004. *The Death of Environmentalism. Global Warming Politics in a Post-Environmental World*. The Breacktrought institute. http://thebreakthrough.org/archive/the_death_of_environmentalism.

Shellenberger, M., and Nordhaus, T. 2010. "Cap and Charade: The Green Jobs Myth." *The New Republic*, October 28. http://www.tnr.com/print/article/politics/78209/clean-energy-jobs-obama.

Shellenberger M., and Nordhaus, T. 2011. "Love Your Monsters: Postenvironmentalism and the Anthropocene." *The Breacktrought Institute*. http://thebreakthrough.org/index.php/programs/philosophy/love-your-monsters-ebook.

Shelton, T., Zook, M., and Wiig, A. 2015. "The 'actually existing smart city'." *Cambridge Journal of Regions, Economy and Society* (forth. 2015).

Shiva, V. 1993. *Monocultures of the Mind. Perspectives on Biodiversity and Biotechnology*. New York: Palgrave MacMillan.

Söderström, O., Paasche, T., and Klauser, F. 2014. "Smart Cities as Corporate Storytelling." *City* 18(3): 307–320.

Soja, E. 1989. *Postmodern Geographies: The Reassertion of Space in Critical Social Theory*. London (NY): Verso.

Soja, E. 2010. *Seeking Spatial Justice*. Minneapolis (MN): University of Minnesota Press.

Spahl, T. 2014. "'Environmentalism Has Become a Religion'." Accessed May 19, 2014. http://www.spiked-online.com/newsite/article/environmentalism-has-become-a-religion/15033#.VYEksaYTjaY.

Speth, G.J. 2008."Environmental Failure: A Case for a New Green Politics." *Environment 360*, October 20.

Stanforth, C. 2006. "Using Actor-Network Theory to Analyze E-Government Implementation in Developing Countries." *Information Technologies and International Development* 3: 35–60.

Stengers, I. 1997. *Power and Invention: Situating Science*. Minneapolis: University of Minnesota Press.

Stengers, I. 2005. "The Cosmopolitical Proposal." In *Making Things Public-Atmosphere of Democracy* edited by B. Latour and P. Weibel. Cambridge (MA): MIT Press.

Susan Leigh Star. 1991. "Distribution of Power: Power, Technologies and the Phenomenology of Conventions. On Being Allergic to Onions." In *A sociology of Monsters: Essays on Power, Technology and Domination* edited by J. Law, 26–56. New York/London: Routledge.

Tansley, A. 1935. "The Use and Abuse of Vegetational Concepts and Terms." *Ecology* n 16.

Taylor, C. 1991. *The Malaise of Modernity*. Concord (ON): Anansi.

Taylor, C. 1992. *The Ethic of Authenticity*. Harvard: Harvard University Press.

Taylor, B. 2013. "Resistance: Do the Ends Justify the Means?." In *Is Sustainability Still Possible?* edited by Worldwatch Institute (Linda Stark, ed.), State of the World 2013, 304–316, 421–423. Washington (DC): Island Press.

Taylor S., Zook, M., and Wiig, A. 2015. "The 'Actually Existing Smart City'." *Cambridge Journal of Regions, Economy and Society* 8: 13–25.

TEEB. 2012. *The Economics of Ecosystems and Biodiversity in Business and Enterprise.* Edited by J. Bishop. London and New York: Earthscan.

Thatcher, M. 1982. *On the Falklands War May 26th 1982.* Accessed June 30, 2009. http://www.totalpolitics.com/quotations/quotations.php?cat_id=179.

Thatcher, M. 1988. *Speech to the Royal Society, September 27th 1988.* Accessed May 1, 2008. http://www.margaretthatcher.org/speeches/displaydocument.asp?docid=107346.

The Nobel Peace Prize. 2007. *Nobelprize.org. Nobel Media AB 2014.* Web. Accessed October 14, 2016. http://www.nobelprize.org/nobel_prizes/peace/laureates/2007/.

Thomas, F, and Homer-Dixon. 1999. *Environment, Scarcity, and Violence.* Princeton (NJ): Princeton University Press.

Tracey, D. 2007. *Guerrilla Gardening: A Manualfesto.* Gabriola Island (BC): New Society Publishers.

Transition Network. 2015. "About Transition Network." https://www.transition network.org/about.

Trudgill, S. 2001. "Psychobiogeography: A Meanings of Nature and Motivations for a Democratized Conservation Ethic." *Journal of Biogeography* n 28.

U.N. 1987. "Report of World Commission on Environment and Development: our common future." Published as Annex to General Assembly document A/42/427.

U.N. 1992. "Rio Declaration on Environment and Development." Conference on Environment and Development. Rio de Janeiro, June 3–14.

U.N. 2002. "Report of the World Summit on Sustainable Development." Johannesburg, South Africa, August 26- September 4.

UN Global Compact and IUCN. 2012. "A Framework for Corporate Action on Biodiversity and Ecosystem Services."

UNEP. 1972. "Declaration of the U.N. Conference on the Human Environment." http://www.unep.org/Documents.Multilingual/Default.asp?documentid=97%26;articleid=1503.

UN-HABITAT. 2010."Cities for All: Bridging the Urban Divide." *State of the World's Cities.* 2010/2011.

Plumwood, V. 2002. *Environmental Culture the Ecological Crisis of Reason.* London: Routledge.

Van Acker, R.C, Brûlé-Babel, A.L., and Friesen, L.F. 2003. "An Environmental Safety Assessment of Roundup Ready®, Wheat: Risks for Direct Seeding Systems." In Western Canada report prepared for the Canadian Wheat Board for submission to Plant Biosafety Office of the Canadian Food Inspection Agency.

Van Haren, Lena. 2015. "Environmental Justice Case Study: The U'wa Struggle Against Occidental Petroleum." http://umich.edu/~snre492/Jones/uwa.htm.

Vandana, S. 1993. *Monocultures of the Mind: Perspectives on Biodiversity and Biotechnology*. London (UK): Zed Books.

Vanolo, A. 2014. "Smartmentality: The Smart City as Disciplinary Strategy." *Urban Studies* 51(5): 883–898.

Wackernagel, M., and Rees, W. 1998. *Our Ecological Footprint: Reducing Human Impact on the Earth*. Gabriola Island (BC, Canada): New Society Publishers.

Wakefield, S., Yeudall, F., Taron, C., Reynolds, J., and Skinner, A. 2007. "Growing Urban Health: Community Gardening in South-east Toronto." *Health Promotion International* 22(2): 92–101.

Weber, M. 2005. *The Protestant Ethic and the Spirit of Capitalism*. London: Routledge (or.ed. 2005).

Wekerle, G.R., Sandberg, L.A., and Gilbert, L. 2009. "Taking a Stand in Exurbia: Environmental Movements to Preserve Nature and Resist Sprawl." In *Environmental Conflicts and Democracy in Canada* edited by L. Adkin, 279–297. Vancouver: UBC Press.

Whatmore, S. 2000. "Heterogeneous Geographies. Reimagining the Space of N/nature." In *Cultural Turns/Geographical Turns: Perspectives on Cultural Geography* edited by I. Cook, D. Crouch, S. Naylor and J. Ryan. Harlow: Pearson.

Whatmore, S. 2002. *Hybrid Geography*. London: Sage Publications.

Whatmore, S. 2003. "Introduction: More than Human Geographies." In *Handbook of Cultural Geography* edited by K. Anserson, M. Domosh, S. Pile and N. Thrift. London: Sage.

White, D., and Wilbert, C. 2006. "Introduction: Technonatural Time–Spaces." *Science as Culture* 15: 95–104.

White D., Rudy, A., and Gareau, B. 2015. *Environments, Nature and Social Theory. Toward a Critical Hybridity*. New York: Palgrave Macmillan.

Winn, M., and Pogutz, S. 2013. "Business, Ecosystems, and Biodiversity: New Horizons for Management Research." *Organization & Environment* 26(2): 203–29.

Wissenburg, M.L.J. 1997. "A Taxonomy of Green Ideas." *Journal of Political Ideologies* 2: 9–50.

WWF. 1995. "Solomon Islands project." http://www.wwfpacific.org/about/solomon_islands_/.

Young, I. 1990. *Post Environmentalism*. London: Belhaven Press.

Young, I.M. 1990. *Justice and the Politics of Difference*. Princeton: Princeton University Press.

INDEX

© The Author(s) 2016
C. Certomà, *Postenvironmentalism*,
DOI 10.1057/978-1-137-50790-7

143

Printed in the United States
By Bookmasters